Thirty years ago, all the La Unión mines closed. The metals didn't run out, but the ever-changing market values and the health problems of the miners were part of the reason that the industry died. This is by no means the whole story. The whole history involves many minor closedowns, fluctuations in the price of metals, other factors like War and, finally, issues of contamination. There are still vast quarries, but these seem to be used, largely, for building stone or dumping. The landscape is littered with ruined mining buildings and thousands of well-like ventilation shafts.

"Washing Amethysts in the Bidet" is more than a comprehensive look at the huge range of minerals that can be collected in this fascinating area in Murcia it also reveals where they can found and what the terrain is like. Fiona has filled the book with her own experiences, real stories about the abandoned mines, the people she has met on the way and fascinating historical anecdotes about the places she visited whilst exploring. This is more than a book about collecting geological specimens, although that aspect is well covered. It is a book about people, mining communities and stories of a way of life that is now gone.

Published in 2023 by

UP PUBLICATIONS

U P Publications, St George's House, George St, Huntingdon, Cambridgeshire, PE29 3GH. UK +44 208 133 0123
manager@uppublications.ltd.uk

Copyrights © 2020, 2023 Fiona Pitt-kethley

Fiona Pitt-kethley has asserted her moral rights under the Copyright, Designs & Patents Act 1988, to be identified as the Author of this Work

All Rights Reserved - No part of this publication may be reproduced or transmitted by any means, electronic, mechanical, photocopy or otherwise, without the prior permission of the author or publisher - except for the purpose of research or editorial use. This is a work of non-fiction based on the information available at the time of publication and the author's experience in the field. A catalogue record for this book is available from the British Library. Image Copyrights: Fiona Pitt-kethley Licensed Fonts: Caviar Dreams, Veracruz Serial, Concept.

ISBN 13: 978-1-912777-36-5 Soft Cover
ISBN 13: 978-1-912777-37-2 eBook

www.uppbooks.com

Washing Amethysts in the Bidet

Fiona Pitt-kethley

2023

Fiona Pitt-kethley

For Alexander who was often a companion down mines.

For James who sometimes drove us to them and put up with having the bidet filled with muddy amethysts from time to time.

My thanks also to friends in the Asociación Cultural Mineralógica de la Sierra de Cartagena-La Unión for excursions and their help, sometimes, in identifying minerals, most especially to Ginés López García, Juan José Martínez Pardo, Eliécer Pérez Sánchez, Fuensanta Alcaraz Saura, Diego Martínez de Ojeda Martínez, José Javier Saura Nadal, Antonio Retamero Tejada, José Luis Guirao Escudero, José De Luis Del Campo and Antonio "El Sevillano" (RIP). To "Robles" a fellow rock-hunter and excellent painter (RIP). Also, thanks to Manuel Morales for information on mines and the Peñarroya booklet. To Professor Javier García del Toro for garum tastings and excellent lectures and to Rogelio Mouzo Pagán (RIP) for his blogs and José Matías Peñas for his films of contamination. To Isabelle Carbonell for company down a mine and for a fine set of waders. To Ana Rama for introducing me to the area I live in now. And also, to Robert Barnes (RIP) and Andrew Wood for being mining buddies, occasionally

Contents

Introduction 9
1: La Unión 24
2: Amethysts and Quartz 39
3: A Little Mining History 51
4: El Descargador 67
5: The FEVE 77
6: Llano del Beal 83
7: Cabezo de Ponce, Peña de Águila and the Campos de Golf 94
8: Opal 102
9: Pyrite 108
10: Galena 115
11: Escombreras 120
12: Lo Campano and San Julián 128
13: Barite 137
14: Monte Miral 141
15: Fluorite 158
16: Zinc Blende 171

17: Portmán 186
18: El Gorguel 191
19: Gypsum 200
20: Quarries, Deviline and Micros 205
21: Lavadero Roberto and Peñarroya 220
22: The Big Bosses, Strikes and Riots 228
23: The Close Down of Mining 242
24: BIC 248
25: The Dying Mar Menor 252
Glossary: Minerals of the Sierra Minera 260
Minerals 261
Micro Minerals 278
Bibliography 287
Index 290

Washing Amethysts in the Bidet

1: Cala Cortina

Introduction

They came in their thousands in the nineteenth century to try their luck in the legendary mines of La Unión in Murcia. Some were Gitanos from Almeria and Andalusia. For all these men and their families there was a new beginning and the chance quite literally to make their fortunes. A few succeeded financially while others became slaves to a hard life, working for a pittance and dying young.

Not all were miners in this golden age, some started in the catering business. Man must eat. They set up tiny cafés and bars where the men could spend their wages. What did they eat? These days, workmen and schoolchildren queue for bocadillos in my local supermarket every weekday, half loaves filled with ham or chorizo. The men often get a large bottle of beer to wash it down. Some also buy empanadillas, pasties filled with frito, a tasty mix of tomatoes, courgettes, onions and peppers.

Modern Spain is almost totally non-vegetarian.

Some say this is a reaction to the hard-up times in the Civil War and earlier when people sometimes survived on beans and vegetables. Miners carried a kind of lidded pot that could perhaps have contained a mess of beans or stew in one half. Spain has a great many recipes for tasty casseroles made with beans or lentils. They often end up as first courses in menus del día in these spoilt times but were probably a family's entire food in the nineteenth century. Packets of chickpeas and other beans feature largely in the basic foods given out to poor families since La Crisis.

The charities concerned are probably unaware that these are not popular items because of the length of time soaking and cooking and that they often get sold on or swapped by the families that receive them.

Poems about mining occasionally talk of the black bread eaten in the mines, not rye bread, which is not popular in Spain, but ordinary bread blackened by dust...

Some entrepreneurs from the neighbouring provinces opened shops where tools and explosives were sold. The men also exchanged information in these places on who wanted workers. Spain abounds in bureaucracy and modern-day shops must go through many hurdles in the way of paperwork before they can open. Probably it was all easier in pre-Franco days. Even now, where I live in Cartagena, there are many small illegal shops. Before the fires of San Juan in June, a host of ladies buy boxes of fireworks and cash-in quietly by selling bangers, small rockets, etcetera, to the local kids. Some fireworks are merely aimed at making the maximum noise. Others are pretty or have funny names like drunken bees. I sometimes feel that the local love of explosives is a hangover from mining and military days. La Unión was a prosperous town in the nineteenth century. It was two villages pulled together to make one entity.

Herrerías, named after the blacksmiths working there, became the poshest part where the nouveau riche mine-owners built splendid properties. El Garbanzal was mainly for the workers.

Roche and Portmán joined La Unión later. Roche is a rural village. It lies close to the motorway. The houses are pricey, but I am told, by those who were born there, that it is no longer a proper village because it has lost some of its soul. Soul is something that La Unión itself has in abundance. As you get to know the place you are conscious of the weight and importance of its mining history and its music. I grew more and more attached to the town as I visited it from the soulless urbanisations of expatland on the Costa Blanca. The same is true of the bleaker village of Portmán, which it annexed in time.

In the days of the Romans, it had been the glorious Portus Magnus where the rich of those days lived in villas with spectacular mosaic floors. In later times, it was a small fishing village. Today it has an uneasy atmosphere. Holiday homes for English golfers have been built close to traditional Spanish houses. The houses are dearer than many on the Costas. I wonder why. Though the backdrop of hills is beautiful, purchasers have been sold a dubious asset. It's a town with few facilities in winter. The centre of the beach also bears terrible marks of pollution. There are many plans for its future, but it still has not managed to get down to the decontamination and improvement that is needed. Things have advanced a bit across the years though. Nature is restoring some bits by itself and a general plan has been evolved.

Tests have proved that a mixture of pig purine and marble dust greatly improves contaminated soil and thereafter, careful planting can restore it further. A German company was scheduled to do part of the recovery work until its head was arrested, accused of fraud.

Some siderite from the contaminated soil was to be sold to China also. Most of the residues were then to be dumped in the San José and Gloria quarries and San Valentín.

I intend to try to dissuade them from the latter course, nearer the time, as water from San Valentín filters back to Portmán via the Túnel José Maestre and then seeps down the hillside. My explorations under and above ground have shown me this. Few other people are aware of it. Besides, the mineral assets of San Valentín are too good to lose. San José and the Gloria quarries are much more limited, though even these are a loss.

How did the prospectors make their way to La Unión? Walking, or riding, across the other sierras that lie along the way? Someday I wish to walk that entire route and get the feel of it. I have seen horses ridden through the Calblanque Park on a road used by walkers, cyclists and quads from the village of Llano del Beal. One Sunday as we were walking, we saw three horsemen make it to the top, their black horses panting and dripping with sweat. It steamed off them in the cold spring air. One of my preaching Welsh ancestors rode over much of England and Wales with Wesley clocking up considerable mileage. Perhaps the prospectors in the nineteenth century rode their horses as hard as the men in Calblanque, never stopping until they found an inn for the night. Perhaps they were more like my grandfather who walked his way at a gentler pace round much of South Wales and Devon, preaching to different communities. Perhaps the men from Almeria and Andalusia took an even slower path with a donkey pulling a cart filled with their possessions.

There was no motorway between Cartagena and Vera then, no tunnels through the hills, just the old winding roads. Why did they not stop at the nearer mining areas in the Sierra Almagrera of Almeria, or by Águilas or Mazarrón? What brought them on inexorably to La Unión? Was La Unión considered to be some sort of Eldorado? I suspect that this was the case. A few went on to become mine-owners and industrialists of considerable wealth. Notably these were foundry workers rather than miners. There are still several palaces that were once owned by some of these men in Cartagena. Some of these were designed by the wonderfully imaginative Catalan architect, Victor Beltrí. He also built the market hall of La Unión. The banks own most of the palaces, these days. They are large elaborate follies, too expensive for modern families. Yet they are some of the most beautiful architecture in Cartagena. No expense was spared in their details, the tiles, the frescoed and stuccoed ceilings, the stained glass, the iron and bronze work. Their style is labelled modernismo in Spain. To the British and French, it is Art Nouveau. A kind of homage is paid to the mining industry in some of its details. This is directly commemorated in bronze work including a relief of a mining castle on the great doors of Casa Cervantes in Calle Mayor. This building is now the Sabadell bank. There is also a less direct form of homage in the widespread use of metal in architecture where one might expect marble, stone, and slate. A great many of the most decorative buildings are roofed in zinc cut into overlapping plates and formed into cupolas, etcetera. They have a rather Parisian feel and have stood the test of time. Zinc is also used in plaques that look like stone reliefs. Inside some of the buildings there are pillars which look like marble, but which are in fact cast iron. Some balconies, stair railings and doors also have elaborate bronze work.

    Cartagena has guided routes that mention in passing the financial origins of these palaces, but little money and effort is put into promoting the mining history and its industrial architecture. Most is just left to rot. One day, I hope the local politicians will do something about this. Cartagena has a huge problem of having a magnificent heritage of archaeological remains and architecture without much money to fund its restoration.

While much of this architecture was funded during the mining boom, some of the later examples took advantage also of its decline after the First World War. The Casa de Misericordia, built in 1929, used a lot of Canadian pine in its construction. The timber was taken from the demolition of mining properties that had recently closed. The Casa de Misericordia was a charitable establishment, but it is now owned by the University.

How and why did I get interested in all this and move to Cartagena to be close to it? Our life's interests can be born in strange ways. My son loved museums, mostly archaeological ones, from an early age. When we had run through all those within a circuit of fifty kilometres I took him to the small mineral museum in La Unión. He was eight at the time. He was not normally prone to nicking things but was riveted by some small pyrites in a working miniature model of mining trucks and was determined to take one home. I tried reason. I tried bribery. But neither worked. There was no way I was going to get him out of there without a pyrite, until, in a moment of inspiration, I said: "We are going up the hills and we are going to find at least ten different specimens as good or better!" Thus, a new hobby was born for both of us.

I had always liked semi-precious stones but only knew them in their polished form. Finding them in their raw rough state is an exciting experience. We went for massive walks across the hills dressed like demented boy scouts and came back with so many rocks in our knapsacks that the whole exercise resembled

SAS training. Sometimes we collapsed into a café covered in mud for a plate of paella on the way home. Fortunately, Spanish restaurants are very forgiving about the sartorial inadequacies of their customers. Eventually we moved to Cartagena, which is on the edge of what was once one of the greatest mining areas of Spain, the Sierra Minera. For years before we had taken fifty-kilometre bus rides at the weekend to spend our Saturdays and Sundays there.

Lead and silver were mined by the Tartessans and traded with the Phoenicians, even before the Carthaginians and Romans came. After the Punic Wars, a cruel tribute was paid of massive amounts of silver worked by slaves. Copper and iron were also mined there. Cartagena was an important area for the manufacture of arms for the Roman armies. The right ores were close by and there were skilled artisans in the city itself. This is something local archaeologists have only talked of recently as more and more Roman remains are uncovered.

As the centuries passed, other metals were discovered: tin, manganese, zinc, cadmium, etcetera. The beautiful mineralisations that had formed alongside the metals were a low priority to the workers and their bosses.

Mining went into decline at the end of the Roman period but there was a little going on judging by the evidence of mining concessions for various minerals. The nineteenth century saw the equivalent of the California Gold Rush. Many workers came to make their fortunes.

A few did in a spectacular way. They left behind them wonderful architectural follies in the towns of Cartagena, La Unión and Portmán. My favourite character from those days was Pió Wandossell, who made several million from nothing, was a prominent freemason and had twenty-three children!

Three of his sons were in the first team that was later known as Real Madrid.

The nineteenth century also spawned a wonderful vein of poetry and song, the Mineras and Cartageneras. Once a year, the town of La Unión hosts a prodigious Flamenco festival, the Cante de las Minas (songs of the mines). While the poems look as discreet as a haiku on the page, they are heart-rending and violent when you hear them sung. This is real Flamenco, not cheap tourist stuff. As with hill-walking and rock collection you need stamina. The final concert starts at 11 and goes on till 4.30 in the morning...

The Flamenco connection in La Unión was largely triggered by the huge influx of Andalusians as workers. They brought their culture with them. There are still a lot of people with Andalusian blood in the area.

Thirty years ago, all the mines closed. The metals didn't run out, but the ever-changing market values and the health problems of the miners were part of the reason that the industry died. But this is by no means the whole story. The whole history involves many minor closedowns, fluctuations in the price of metals, other factors like War and, finally issues of contamination. There are still vast quarries, but these seem to be used largely for building stone or dumping. The landscape is littered with ruined mining buildings and thousands of well-like ventilation shafts. Most of the wells have a wall around, some do not. This makes for dangerous walking conditions when you are off the main roads. You must proceed carefully and slowly. Some hills are natural, others are tips. Many are a combination of waste tipped on to solid rock. This is particularly treacherous under foot. They are best negotiated with the aid of a walking pole. We soon purchased hard hats and started venturing into the safer-looking

2: Market, La Unión

mines, mainly those with straight tunnels hewn from the rock or lined with bricks. Gradually we progressed to tackling ramped caves with climber's rope. There's an extra buzz to getting your minerals straight from the rock face. Occasionally we met other rock-hounds chipping away. We joined a local society, which also takes trips to other parts of Spain. We discovered the Asociación Cultural Mineralogica de Cartagena y La Unión at the annual mineral fair. To join it I had to go and talk to one of the founding members and fill out a bank order. Pepe was a mortgage manager, or similar, in Caja Murcia, so our meeting took place at the bank. I let him do most of the talking, as my Spanish was very imperfect at the time. He asked if we had a collection and I said yes, conscious that it was not yet the sort to be displayed proudly behind glass, but rather, a few hundred dirty stones jumbled in boxes beneath beds.

At first, we lived too far away to go on many trips and just read the annual bulletins and visited the mineral fair. We became regulars on the excursions once we moved to Cartagena. One year, the society's annual bulletin contained the story of the death of one of the members who was killed instantly by a rockfall in a mine in front of his friends. It was a reminder that great care is needed. Sections of rock can stay in one place for a millennium then drop in a moment. The following year at the Feria de Minerales, a silence was devoted to Virginio Moreno. He was a young man with a wife and children. The mineral fair also pays respect to the oldest miner left alive. La Unión does not forget its origins.

Our first proper mineral fair was the large annual one in Holy Week in La Unión. We bought many small minerals. These days we buy much less as we prefer collecting our own. Purchases these days tend to be copies of the Mineral magazine, Bocamina, or boxes and stands to display our own minerals.

The fair in La Unión is of such a high standard we have never dared to rent a stall. It's expensive per square metre, plus you need the ability to take credit cards and everything is done in as professional a way as possible. Through this fair, we have got to know many dealers throughout Spain. We usually go for a group lunch or two with them in a local restaurant. They are unusual and interesting people, especially those who still collect rather than just buying in jewellery, etcetera. Most have as much physical strength as intelligence. I have always admired the equal balance of physical and mental skills and don't feel I have much in common with the sort of academic who is a couch potato. Mineralogists and archaeologists are much more up my street. In our first visits to the fair, son often pointed one or other stall keeper out and whispered: "He's a miner, just like us." There's a kind of perceived solidarity with those who still go down mines or hack rocks from quarries rather than with those who buy in a lot of ready-polished stones from China via eBay.

One interesting dealer and collector we met was José Vicente Casado of Leon. He has the fascinating profession of "cazameteoritos" (meteorite-hunter) and has 400 of them in his home. He sells specimens of these and other rocks he has collected around the world and has given a large valuable meteor to the Valencia Museum. When a meteor falls in an area where he has a chance of finding rocks, he is off across the world. Deserts give the best chance of finding pieces. A fall over a forest would present considerable difficulties by comparison.

When we first started, we swapped some of our minerals at a monthly fair in San Vicente del Raspeig, near Alicante. Someday, I would like to see Cartagena set up something similar. The collectors there liked the rough yellowish opal and sparkling hemimorphite, rare in their region, so my son struck good deals for trilobites and bottles of mercury, etcetera, in exchange. We usually sold our rough specimens dirt cheap. Sometimes other collectors bought ones they thought had potential for improvement.

We began to realise then that we were collecting too much rubbish and not presenting our rocks well. Specimens of quartz, for instance, need most of their points intact and not too much staining.

Some dealers are fantastically good at the cleaning side. Our friend, Eliecér is one of the most professional at this. Some others are cruder in their techniques and leave stones in acid far too long till they become fragile. A few other minerals have special techniques. Gypsum is too soluble for water and needs careful brushing. Acicular aragonite is so delicate you can only use water on it through a tiny syringe. We use my husband's diabetic syringes for this.

Our collection has now become so massive that we have rockeries in every room. Some stones arrive covered in mud and need to be cleansed. Water and washing-up liquid cope with basic dirt. Toothbrushes also come in handy, electrical or otherwise. These can be used with toothpaste or Mistol, a cheap brand of washing up liquid. Ultrasonic cleaners can also be used on small pieces. Layers of calcite and some ingrained dirt may need hydrochloric acid. Fortunately, that is cheap and readily available in Spanish supermarkets. The fumes are bad, so you need lots of protection.

The stones are then soaked in water to remove the residues of acid. I am now also experimenting with vinegar which works in a similar way to hydrochloric acid but is much slower.

There is an art to using acid that I have not wholly

mastered. The most professional mineral dealers have. I am still learning. Some pieces like barite are not harmed by a long soaking. The piece is then left in water for three times the length of the time it has spent in acid. The water should be changed from time to time. Quartz that has calcite that needs to be removed can be treated roughly also. Too much soaking seems to make quartz less attractive though. It can become fragile and for me, the soul is in some way lost. Less experienced dealers sometimes make the mistake of overdoing it. Other minerals may require subtler treatment. A tiny amount of acid brushed on to zinc blende may improve its appearance. Too much, dissolves it, releasing stinking hydrogen sulphide. After a year or two we gave up San Vicente del Raspeig as it became too rulebound.

These days I sell rocks for about eleven days in August outside the Cante de las Minas concerts. Most of the customers are not collectors but concertgoers looking for souvenirs of the mining area. I also do a yearly collectors' fair in San Pedro and will try for some other spots. I am not quite a professional dealer, but I make a little extra cash and it gives me the chance to enthuse about collecting with anyone willing to listen.

The real professionals, of which there are more than a few amongst my friends, are willing to clock up immense mileage going to the mineral fairs around Spain and Europe. There's plenty to collect in Spain, but Spain doesn't pay as well for minerals as fairs in Germany for instance. Within Spain, Valencia, Madrid, and Barcelona can charge higher rates for specimens than La Unión despite the mineral history of that spot.

Dealers are mostly couples who can take turns with the driving. The women are usually very good-looking and often concentrate on the jewellery end of the business. Most of the obsessive rock gatherers are men. Spanish friends tell me that decades ago there was much more travelling around for work. Some families shifted themselves from harvest to harvest in Spain and France. The concept of driving a van full of rocks through Europe to a specific mineral fair is not so far different.

Washing Amethysts in the Bidet

3: *Tunnel interior, La Unión*

1 : La Unión

When we first started to visit the area, we usually got a bus to La Unión. We set off early from Playa Flamenca close to Torrevieja where we were living at the time. When we reached our destination, we were ready for a second breakfast like hobbits. We usually shared a large piece of tortilla in Bar Minero. I had either a beer or a coffee on the side while my son sunk a cola cao, a hot chocolate, to keep him warm on winter days.

    We also consulted a huge map on the wall, which showed eleven hundred mines. I now have a copy of this map, which came free with the magazine Bocamina in the edition covering La Unión. The plan was put together by the engineer, Carlos Lanzarote, in 1907.

    It is the best all-round source for the position and names of mines in the district. The La Unión edition of Bocamina has long ago sold out. Describing it as La Unión rather than the Sierra Minera or the Sierra de Cartagena is fuelling a common misconception. A major proportion of the places described are on Cartagena's soil.

    Years later, I was to meet and become friends with Manuel Morales who had supplied much of the information in the article. He has a mineral shop in Cartagena and has gathered fabulous specimens across the years as well as collecting other mining memorabilia by way of papers, mining lamps and documents.

After a foray to one or more of the local mines we were ready to get a bus home again.

By then, I was usually carrying 20 or 30 kilos of rocks on my back as my son had become too tired to carry anything. The bus stop was at Café Llamusi, which now has another name and another owner. Llamusi was a local religious poet and his psalm-like poems were displayed on the walls. The young owner was religious also. He liked children and always presented my son and others with free bags of crisps to take away. He gave my son so many bags my son started to hate crisps, which was probably a good thing. When my son thanked him, he would usually say not to thank him, but to thank God.

As the bus left La Unión, I always felt some sadness and watched the last few kilometres of the Sierra Minera with love. La Unión, for all its faults, is a town with a soul. The expat area I lived in was all newly-built and lacking history.

When I visited La Unión, or Cartagena I started to buy books on the area. Some were by the local artist and writer Asensio Sáez. His style was a little too purple for my taste but there were interesting facts in his writings. Others were technical books produced by the Fundación Sierra Minera. These contained useful plans and pictures, which enabled me to identify individual mines and plan routes more effectively. When we moved to Cartagena years later, we had also looked at La Unión as a possible location.

I love both places in different ways.

La Unión suffers much graver unemployment problems than Cartagena. It currently stands at 49 per cent. It has never entirely recovered from the end of mining.

4: La Unión Town Hall

There are still some fine bits of architecture there from richer times. The Casa del Piñon now houses the town hall. It was taken over after the previous owner left it in a state of poor repair. It's a huge modernist building, with spiral staircases and metal covered cupolas, balconies, and pilasters. It was built in 1899, by Pedro Cerdán Martínez for the mining impresario, Joaquín Peñalver Nieto. The name Piñon was a nickname as he was very short. The building was commissioned as an investment. The apartments inside it were to be let to rich families.

The market hall is also sometimes known as the Catedral de Cante, the cathedral of song, because of its use for Flamenco concerts. It's a beautiful building designed by the Catalan architect, Victor Beltrí. Decorations on the outside still contain motifs like fishes from the days when it really was a market. Inside, there's a spectacular use of cast iron in the style of the Eiffel tower. Birds often hang out in the glass and girder roof. Apart from the Flamenco festival and the huge mineral fair just before Easter it is also used for other cultural events.

Shortly after we had begun collecting, my reading about La Unión made me aware of the form of Flamenco it had produced, the Mineras.

I was painfully ignorant about Flamenco then and perceived it as largely a dance thing with women in long flouncy dresses.

Out of curiosity, when my son was 8, I bought us tickets for the final of the Cante de las Minas festival. I wanted to hear the songs the miners sang down the mines we had begun to visit regularly. I didn't quite realise what I was getting into as the concert ran from 11 to 4.30 with an interval in the small hours. Son passed out round about 2 and slept quietly leaning against me.

At 2.30, in the interval we had churros and chocolate. These are disgustingly rich deep-fried doughnut like strips dipped in thick chocolate. I dislike them normally, but they are effective in keeping you awake in a situation like that. The indigestion and extreme calories help.

I was riveted from the first by the sound of the mining songs. These always begin with a kind of cry of pain: "aa-ee" etcetera. I became hooked on them and now have many CDs. I have been to literally hundreds of Flamenco concerts, many of which have been free in Cartagena. I retain a special affection for the mineras and their link to the land I have grown to love. The words are usually in five-line stanzas and often only one verse long.

They give scope for huge emotion and improvisation. Most of the extant verses are anonymous. I have a little book of them collected by the famous singer, Pencho Cros. He was a great exponent of them and lived a very humble down-to-earth life. He refused to tour as a singer, being unwilling to leave his hometown. From mining he graduated into industry and eventually lost some fingers in an accident.

Mineras are very poetic in a chilling kind of way. They concentrate on the tragic aspects of the life and its hardness down the mines. There is nothing about hitting a rich vein of metal only the horrors of darkness and closeness to death. Probably the nearest thing to them in English literature is Nash's Timor mortis conturbat me. Here is a three stanza one:

> *De las minas no me quejo*
> *porque nunca me fue mal,*
> *pero ahora me las dejo*
> *porque quiero descansar,*
> *ya que me encuentro muy viejo.*

Compadre si va usted al cielo
hágame usted este favor:
pregúntele usted a mi abuelo
dónde se dejó el legón
y el capacico terrero.

No se espante usted, señora,
que es un minero quien canta.
Con el "jumo" de las minas
tiene ronca la garganta.

I have translated this very loosely into four-line stanzas as these seem more English and English is a more concise language than Spanish.

I won't complain about the mines.
They never did me harm. It's time
To quit the job. I need to rest.
I feel so old. I'm past my best.

If you should go to heaven first, friend,
I have a message I would send:
Grandad, where did you stash your tools,
The earth-stained bag and the pick you used?

Lady, don't shrink back in fear
From the singing miner here.
The rattle in my throat you hear
Comes from the fumes I breathed down there.

The mineras are reckoned to be the closest amongst Flamenco forms to Arabic music. Their form is nearest to the Fandangos of the area and Tarantos. It was evolved by Rojo El Alpargatero. As his nickname would suggest, he was a red-haired espadrille-maker who branched out into singing in the café cantante.

His real name was Antonio Grau Mora. He was born in 1847 in Callosa de Segura. After time spent in Malaga, he moved to La Unión in 1882 with his pregnant girlfriend. At this period, he was already an established singer. The basic form of the minera seems to have existed amongst the miners from the mid-nineteenth century, but he sometimes gets the credit for inventing it rather than shaping it and refining it.

His son, Antonio Grau Dauset toured Europe singing them and even worked in Russia. The form fell out of favour for a while after that and was revived in the 1950s. I sometimes close my eyes when I hear one. For me the linear nature of the tunes resembles exploration of a mine. In my mind, I can feel myself traversing a long passage before taking a few steps either up or down to another. Mines are more organic than architect-designed, and their form is dependent on where the minerals are. The levels are not as easily viewed as those in a conventional house.

There are many sung Flamenco forms or palos. If they were drugs, a minera would be a downer and an alegría would be an upper. I have heard many different singers perform them. On the whole, they are better suited to a man's voice as the subject matter is generally directly concerned with mining. On one occasion, I heard a small child sing one at the end of a concert outside my local fish restaurant in Santa Lucia. Didn't think it would work, but the miracle was that the young boy had an old man's voice when he opened his mouth. A shiver ran down my spine.

The first café cantante opened in Cartagena in the Puertos de Murcia. It was known as the Café Cantante del Sol. You could hear Flamenco there in the years from 1880 to 1886. There were also quadrilles and can-can dancers. There was another café cantante in the Plaza de Rey in 1872. Others opened soon after in the same square. Plaza de Rey still has some cafés, but no-one sings there. It is a shadow of its former self thanks to an ill-advised bit of town-planning.

Its palm trees have been replaced with ugly metal sculptures which never shed leaves.

La Unión soon followed suit and opened similar cafés, mainly in Calle Mayor. There is nothing similar left there now, though one of these former cafés has a plaque. The building is now empty and decaying. In Cartagena, Mr. Witt was open for a while and the Carrots Café still has some musical performances. In the old days, it was a free accompaniment to the booze. and a more or less nightly event. In those days, good musicians could not command such high fees unless they were playing and singing for high society. While the Cante de las Minas is on, some marquees in the square outside imitate the old days with performances alongside food, usually after the main concert has ended. The Casa del Folclore in La Palma, a few kilometres away has some of the same atmosphere and puts on nights of music in March to celebrate the fiestas of Santa Florentina.

After seeing our first long Flamenco concert we were adrift in La Unión for a couple of hours before a first bus to the sea at Cabo de Palos. We chatted to the drunks on the square outside the Catedral de Cante, formerly the market hall. Some were singing their own versions of mineras. Another explained to my son how to turn his eyelids inside out. I enjoyed it all, conscious that I did not have to be a politically correct mother now I lived in Spain.

Apart from the Catedral de Cante and Casa del Piñon, there are other small likeable bits of architecture scattered through the town, a smaller house by Beltrí, now a bank, and others by lesser-known architects are worth a photo. The old poor school was turned into a mineral museum for years.

The museum has now been housed in the town hall, while the school was still used for meetings.

There are plans to rebuild it as a museum again but these have stalled due to the finding of large amounts of contamination in the soil. The abandoned industrial buildings of the Maquinistas de Levante close to the railway has been turned into a storehouse for such examples of mining machinery and waggons as could be salvaged from the surrounding countryside.

A few gardeners round the mining areas have been very creative with sculptures made of old mining tools, or carts combined with lumps of quartz and other minerals. Gradually the mining heritage is being stripped and looted.

Most of the old cables and metalwork is ending up in the scrapyards. While the hills still seem to have an unending supply of minerals, it is sad that the old mining buildings are all falling into decay and being stripped, bit by bit.

In an ideological way, La Unión is stripping a part of Cartagena's history. Cartagena has only itself to blame. More and more the mines are being perceived by collectors, the world over, as belonging to La Unión. More than half of them are on Cartagena's soil.

La Unión, with its smaller population and its high unemployment chooses to celebrate the people's history more than the grand architecture and politics and history chosen by Cartagena. Most of the palaces built with mining money are in Cartagena and these are celebrated with only the slightest mention of the men whose lives where shortened making that money for their bosses.

I am conscious also of the duality of riches and poverty at the Cante de las Minas concerts. Ticket prices have risen across the years, although two free concerts at the beginning still allow access to all.

The highest priced concerts are social occasions with tourists from as far away as Japan enjoying the music. Most people are well-dressed for the occasion. Some of the less well-off locals, on the dole, drift into the square to enjoy the atmosphere and listen to the outside café music or make music of their own. The two extremes are gathered together. The boss class and the poor being screwed into the ground. It's an uneasy mix in some ways but the love of music unites them.

La Unión was the site of an unusual murder on the 27th March 1982. It is still unsolved. It was still very much a mining town at the period. The hotel where it happened is unsaleable due to its haunted reputation.

I thought it might make an interesting place to visit and went armed with a torch and my camera to take photos of the graffiti inside. I had already seen some photos and video footage on YouTube so I thought I knew exactly what to expect. My husband promised to stand by and protect me from unseen enemies, if there should be any.

"El Cónsul", Alfonso Martínez Saura, was last seen on the morning of the murder by one of his waiters. That same waiter called the police that evening when he found the hotel shut up. Alfonso was a colourful character.

He started life as a shepherd before going abroad and coming back rich from the Ivory Coast. He used these riches to build a hotel with a tennis court and pool alongside. It opened in 1977. He was an eccentric who sometimes dressed in long African robes which was how he gained the name El Cónsul. He was 65 when he died.

The police broke into the hotel and found the body of Alfonso by the bar with sixty-three stab wounds. He was lying in a pool of blood. Mysteriously every window and door of the place was locked from the inside.

Assuming none were missed out, one possible explanation of this is that Alfonso took a long time to die from the wounds, none of which were in vital organs, and locked his murderer out while he was dying from loss of blood. His wallet was gone but was found on a road two days later with money and papers intact. He had some hair in his hand, but its owner was never found. The weapon that killed him was also never found.

He was separated from his French wife, Gilberte Florentine. Could he have been gay? Perhaps the exotic costumes and the separation could have indicated that. Perhaps the murderer was a chance encounter, a gay lover. Or perhaps it was someone looting valuable property he had brought back from Africa. If this were a novel of a hundred years ago, it would probably be a revenge attack from the followers of an African God whose jewels or image he had stolen. Probably the murder will never be solved unless someone confesses.

Some parts of Alfonso's life were extremely mysterious. He was supposed to have been an honorary consul on the Ivory Coast. But does the consular service ever employ shepherd boys as such? I doubt it. I still see the diplomatic services as being full of snobbery and valuing education and people being of the right family. Probably this was even more true, a few decades ago. Was his money made in some other way then? Possibly. He also had businesses in America. The hotel is said to have been built on the spot where he was a shepherd as a boy.

In 1984, the hotel was closed officially. In 1988, one of Alfonso's sons, a French national, sold the hotel and its land to a company in Torre Pacheco. It has not been occupied since. It is slightly off the beaten track, in the area of La Unión known as Los Camachos and close to the botanical garden of Huerto Pío. It's not the best spot for a hotel in terms of access to the town. It might have had appeal though for those that wanted an out of the way encounter. The hotel is built into a small volcanic outcrop. It is surrounded with a wire fence now and the windows have been blocked. Wire fences almost always have gaps though which provide ways in for curious persons.

In the old photos, I had seen, the windows were not blocked. When I entered, there was therefore less visibility than I had expected. The circular front room of the hotel had been adorned with Caspar the Ghost cartoons between its blocked windows. Most of the graffiti in pictures were still there, such as those claiming that death was present and urging you to leave while you still could. I photographed these and others, one even in Latin, which is unusual. Lovers had left their names. Inevitably someone had done a pentacle and 666. Some areas were so dark I couldn't see much with the light of a small torch, so I photographed these walls with my flash. When I got home, I was surprised to find the first half of my surname with Moriras (you will die) written in red on one of these walls. On the surface, there is no rational explanation of this. It is not a typical Spanish surname. I had told no-one I was going there, so nobody was playing a joke on me.

I didn't feel all that creeped out inside the abandoned hotel which is probably just as well as my husband got bored and deserted me and went off to read a book. I also ate some wild asparagus I found in the inside courtyard.

The inside courtyard provides one possible solution to the locked door scenario. If one of the inner doors into the courtyard was not locked there would have been a way out for someone athletic who could pull themselves up on to the tiles and leave by way of dropping off the roof at one of the sides.

The side that is built into the hill would be the easiest point to leave at. I have seen short films and blogs on the subject but nobody else seems to have thought of this.

Hard to tell in its wrecked state, but my feeling is that it was not a particularly gracious hotel. There's a certain crudeness about the remains of tiling and the colours on the walls. Most of the bedrooms were small. The only part with any style was the long interior patio.

I recently saw a small part of a national TV programme devoted to the mystery. The programme makers inside seemed more nervous than I had been, but did not come up with any very definite evidence with their recordings and photographs.

I noticed that the blood red graffito of my name with Moriras was still on the wall. I felt a strange kind of relief at seeing this. It was obviously a permanent feature rather than something aimed at me personally. There must be a rational explanation. There probably is someone else it refers to who is either dead or alive at this point in time.

A man interviewed on the programme spoke of it being an area where suspicious persons were seen, prostitutes, etcetera. Maybe this was so, back in 1982. These days it just seems to be part of the countryside. One interesting thing did strike me. An aerial shot revealed that the hotel was built in the shape of a keyhole. The circular room is clearly visible from the road. The rest of the building, with its interior courtyard, provides the other part of the keyhole shape. Part of the lower storey is built into the hill, which is strange. I looked back at one of the old blogs on this subject and found that an anonymous poster claimed to have gone for a couple of beers in the hotel.

He was made to feel unwelcome, as if the proprietor couldn't wait for him to leave. He suggested the place might have been a secret brothel. The only surviving photo of Alfonso has him in a suit and bow tie rather than the African robes he normally wore. He looks cynical, jaded and debauched.

Some of those who visited managed to record voices or images, others did not, Perhaps the place is not truly haunted these days, even if it was the scene of exceptional violence, and even if such things as ghosts exist. Will the murder ever be solved? I doubt it. Probably some evidence was missed in a small town with few police officers. What happened to Alfonso's family? Are they still alive now?

I suspect the hotel will never be sold. It's not a good position by today's standards and La Unión is not a rich town. It will probably just fall further into ruin, visited by the curious and adorned with more and more graffiti.

5: *Washing Amethysts in the Bidet*

2: Amethysts and Quartz

I am washing amethysts in the bidet. It's as good a place as any. I found them about ten kilometres from Cartagena, where I now live. The mud flows off them and I begin to see their true colours. These are not dark like Brazilian amethysts. They are pale with more of a concentration of purple near the point. This is the typical local coloration. As I learn more about rocks, I find out how individual they are and how much their precise form is affected by the places they come from.

My husband is not happy when he comes into the bathroom and sees the bidet full of muddy rocks. I think he is about to remind me that none of his previous girlfriends have ever done anything like this.

The colour purple has always meant something to me. When I was seven, I wanted a purple velvet dress to wear on the beach. Naturally, my mother refused. When I was twelve, she made me one of corduroy in a colour called African violet and I wore it for years. I was drawn to certain shades of purple of the bluer kind. There has always been purple in my life. I grew up in a large Victorian house with a rambling garden, at the end of which grew brambles, woody nightshade and a Victoria plum tree. My mother made huge quantities of jam and blackberry and apple tarts. The house often smelt of simmering bramble jelly. Later, in Hastings, I collected sloes for gin and had my own small Turkey fig tree. I brought it back from the garden centre on the back of my bike. Burnham Wood come to Dunsinane. In Spain, I avidly ate the same sort of figs, which grow all over the Sierra Minera. The trees are often blocking the entrances to disused mines or hanging out of the mining wells.

The secret of a good crop, Pliny says, is to restrict the roots with stones. These trees have planted themselves in stony restricted situations. Their figs are remarkably sweet. Some trees even crop twice. The early figs are known as brevas. The local pigeons gorge themselves on them and grow very fat-breasted as a result.

When I was living in Hastings as a young adult, I bought amethyst ear-rings and an old necklace where the stones were joined with pinchbeck links. At this stage I had never imagined I would ever live in a place where I could gather my own. That seemed like the stuff of fairy tales. I didn't even know they were found in Europe.

I am passionately fond of amethysts. My son finds it funny that I can hardly bear to sell them and pick up even the worst specimens with damaged points. I keep a couple of amethysts under my pillow and sometimes clutch them while I sleep. I am not sure if they affect my dreams. They are both fairly damaged pieces but ones that have meaning for me. One is a large biterminate amethyst, slightly smoky in colour. It was in a better state when I found it, but it has been dropped and the points at each end are now blunted. The other I found in the midst of a block of quartz that I split open. It was loose with half formed points and angles and a slight milkiness. The biterminate amethyst was special to me because it was one of the first pieces I found. I acquired it on a day when I didn't get much else. I suddenly saw a faint glitter. It was trampled into a muddy path. I dug it out and brought it home. The second has an aura of mystery for me because of its strange geometry and the fact that it was trapped within another rock.

The geometry is one of the things that attracts me about minerals. Those who do not collect them find it incredibly hard to understand that these complicated forms occur naturally. They conceive of everything that comes out of nature as irregular, which is far from the truth. People who have seen parts of my collection often argue with me and keep asking "Who cut them?" Probably the prevalence of tumbled stones with their smooth surfaces has contributed to these misunderstandings. While non-collectors can accept that a flower might have a fixed number of petals, they will rarely believe that certain types of stones occur in geometric forms with a specific number of sides

Amethyst is, of course, a kind of quartz. A stretch of the hills in the Sierra Minera between Llano del Beal and La Unión is almost entirely made of quartz. Every fold of these hills has its own character and there are many mines and caves. I fell in love with this countryside with its impressive ruins of industrial architecture years before I moved to Cartagena. Most of these hills are overgrown with wild thyme, rosemary, lavender, huge agave plants and the wild fig trees with their deliciously sweet purple fruit. There's also a spectacular array of wildlife: lizards, snakes, hares, frogs, lizards, toads, strange insects like mantises and even blue bees. Occasionally we see eagles circling overhead.

The thin layer of earth and loose stones is perfect for herbs and woody plants like rock roses. There are different fragrances in different areas. Some parts are bitter-smelling with wormwood and rue. Others are fragrant with lavender, rosemary and thyme. It is easy to find attractive pieces on these hills with lots of intact quartz points. Their points have six sides. Sometimes these are crystal clear. Others look like citrine but are more usually just stained yellow from contamination. True citrine is rare but occasionally found. It always has a greenish colour in the metallic matrix behind the crystals.

The countryside must have looked different when there was more water.

There are still many tracks of dried-up streams. They can be used with care as paths up and down the hills. Sometimes they are like staircases with occasional large jumps in level where tiny waterfalls must once have been. There is still some water in the area but much of it is underground, filtering down from the hills above. In mining days, this was sometimes the cause of accidents.

There are still several tunnel mines where water flows out freely.

It was in a dried-up streambed at the start of this area, just outside La Unión, that I found a cluster of smoky quartz points. This was at the start of my mining hobby. I became conscious then that some element of luck or divination plus sharp sight enabled you to find good specimens. Knowledge also helps. Much of what I picked up at first was rubbish. A mineral becomes much easier to find when you have seen someone else's specimen. The pictures in books are not good enough. Some people seem to have the finding gift, others do not. One friend of my son always gravitates to the poorer specimens. He is solely bent on finding something valuable commercially rather than being moved by their beauty first as many collectors are. I know many collectors throughout Spain now and some are rich. While they are not averse to a good profit, their collecting is not intrinsically motivated by it. Always, there is an urge to acquire better and better forms of the minerals they have collected, a search for a Platonic ideal.

Citrine is also known as burnt amethyst. My son persuaded me to sacrifice one or two small specimens in trying to make some. We put a couple of small amethysts in the fire at home. It didn't work. Their colour turned milky instead and they became incredibly fragile. There is a great deal of fragile quartz in the slag heaps that has passed through fires of one kind or another and is completely ruined.

In the quarry known as Corta Emilia there used to be an area where rare green quartz, or prasiolite, was found until the proprietors used it as a dumping ground for inert residues and any worthwhile minerals got hidden. I would love to have a good specimen of it and occasionally think I have found a piece on the section of hillside where there are most amethysts. Invariably it looks less green out of the sunlight when I get it home and wash it. I have some tiny crystals. Eventually I will acquire something better.

It is easy to find loose pieces at your feet amongst the plants on these hills. Quartz of all kinds seems to be commonest in the areas that are covered in thyme. Sometimes people scratch away a little more earth with an all-purpose tool known as a legón. This is a kind of heavy transverse hoe with a handle about a metre long. It has been of use through the centuries in mining and agriculture. It's not as heavy or as forceful as a pick but it is much easier to carry around. I found an old rusty one in the hills and commandeered it. Tools turn up occasionally in this way when you need them.

Gradually, we got to know the most easily visited mines within the area between La Unión and Llano del Beal. Some are tiny, others have long tunnels. The easiest way to reach these mines is via a path that leads off the road that runs to Portmán from Llano del Beal. It's on the right side and very close to the village. It runs past a good fig tree which is so well-known it is hard to get to the ripe figs before anyone else Shortly after you pass that tree on the left you start to notice broken specimens of quartz in the road. There are also many mining ruins along this route, such as the headframes of Mina Brigida and Mina Santa Eduvigis, in Barranco Bilbao. The latter has a wheel system supported on masonry and is seven metres high. Mina Mendigorría, Mina Mentor, Minados La Pobrecita also have some remains in this area. The hillside above is full of quartz specimens.

There are impressive remains of Mina Catón in the next fold of the hillside, el paraje de la finca del Pajarillo. I did find an old entry, which consisted of a small door to a tiny spiral ramp. I spiralled my way down a few turns on my backside until I got to a space so small only someone of baby-size could progress further. It is not so much a mine to enter as one to admire for the old wheelhouses. It's a good area to take photographs. Sadly, some of its metalwork has been stolen. There are several buildings including a washery and a machine house. Some machinery is still in place. A kind person has left a chunky rope so you can easily pull yourself up a slagheap to the area with mining remains. Mina La Pagana also has some ruins in this area. Most of the mines in this part were owned by the Sociedad San Fulgencio. One of the main owners of this society was Hilarión Roux, the French banker and magnate, while Miguel Zapata Sáez had a minority shareholding. Hilarión Roux was also the founder of the larger Companie Française de Mines et Usines de Escombrera-Bleyberg, which was known as Escombrera-Bleyberg for short. It had mining interests in several countries, amongst these the concessions of

6: Mina Catón

the Sociedad San Fulgencio. The mines of this area were all big producers of lead. In 1912, Escombrera-Bleyberg was absorbed by the Société Minière et Métallurgique de Peñarroya, one of the most important Spanish mining companies, founded with French capital from the Rothschild family. The final takeover of some of the subsidiary companies seems to have been slow, as the mines of this area were still registered as belonging to the Sociedad San Fulgencio.

In summer we were grateful for the cold of the mine known either as Los Pajaritos/Los Pajarillos or María Dolores. You need a jacket in there even in August when the temperatures outside are breath-taking. It became a yearly ritual to take a tub of Häagen-Dazs in there and eat it, as it was the one place in Spain where ice cream did not melt in August. María Dolores is a long straight tunnel that runs about half a kilometre into the hillside. The Romans were in this area and it is one mine that probably dates from their time. The Romans liked straight mines as well as straight roads. Several turnings off it peter out quickly. The first on the left has impressive aragonite walls. and leads to an entry to a lower level. Just inside the entrance to the main tunnel there is a ventilation shaft, which may be responsible for the cold inside. Further along, on the right, there is a kind of reservoir of water in a side passage. It is easier to collect pieces outside on the hillside, but there are one or two attractive formations on the walls. One reminded me of a monkey's head in amethyst. Some people find this mine a little depressing because of the cold.

Alongside María Dolores there are various abandoned buildings, the old washeries belonging to the mine. Steps lead up on to the hillside above. The proper path runs out on the hills after these mines, but it is possible to work your way along in and out of the folds of the hills. A silvery mound conceals the entrance to Los Aragonitos also known as San Luciano. This passage is more or less blocked with brushwood at the moment but I have been inside. A passage of some 200 metres leads straight into the hill. At the end there is a turning with some chunks of fallen aragonite on the floor. Along the way you can see some quartz in the walls. It's not the easiest mine to view, as the tunnel is so low you have to pass through it bent double. When I first visited it, my son was small enough to be able to walk upright along it. A little pile of stones which looks like a chimney marks the position of the mine.

These mines ceased working temporarily in 1968. From the 50s they were part of the Minera Celdrán empire. The hillside containing these mines is known as Collado de las Tinajas and the ravine is Barranco de los Pajarillos. There have been various archaeological sites in this area connected with Roman mining. There's a cave house in the area too and fragments of paving and ceramics have been found.

Higher up the slopes in another fold we came across a mine whose name no-one seems to know. The first time we went there we saw three men emerging covered in mud, carrying ropes and some impressive lumps of quartz. They were strangers to us then, but we were to become good friends with two of them in the years that followed.

The mineral world is small.

The passage into the mine led to an impressive cavern. Training our lights into the bottom of it we thought it was probably not advisable to venture down. My son would have done it but for my caution. He was young and light and I could have hauled him out with a rope. But if I had gone down the reverse would not have been possible.

Caution won.

My son reproached me afterwards for not allowing him down. When we next visited this mine, months later, the rocks had broken further and it was no longer possible by any standard. This mine is now known to us as The Chicken Mine because I chickened out. It is rather a special spot. We eventually learned that it has a solar alignment.

Every year, on the 12th of the 12th at 12, light comes through a small aperture in the top of the domed cavern and illuminates a part of the interior. Solar alignments are more often considered a feature of the winter solstice.

I was puzzled at the 12th.

At first, I wondered if it was anything to do with St. Lucy the saint of light whose day is celebrated on the 13th. Or perhaps it is due to the fact that the solstices have shifted in date with the missing days to adjust the calendar. One year, I went inside to see the alignment. It was less dramatic than I expected. I sat in the dark to see it better. I took several photos and the light appeared more in these than in reality. Maybe at one time this light fell on an actual formation of quartz in which case it would have been splendid and dramatic. The stones from there are generally quartz covered with a film of calcite. Perhaps the ray of light at one time descended on a figure-like stalagmite. I would like to think so. I have seen a figure of this nature, without the solar alignment, in the Corycian Cave in Greece.

On the opposite fold of the hills in the direction of La Unión lies Mina Rómulo, a favourite of some of the miners of the Alicante club.

If I failed him on the chicken mine I did at least win brownie points with my son there. We had been eating a sandwich on the hillside outside and I had allowed him use of my knife to whittle a stick. He cut his hand.

Instead of rushing him off for medical help I clapped a cobweb on it and took him down the mine. He realised I had some special qualities that separated me out from all the other mothers at that point in time.

Close to Rómulo is a mine known as León Negro, where a worker of 19 was killed in an accident in 1927. I have found no separate entrance for this. Perhaps the mines interconnected.

7: *View of Calblanque and Cabo de Palos.*

3: A Little Mining History

Cabezo Rajao and the mining museum of La Unión and the National Archaeology Museum in Cartagena provide the best overall view of the mining history of the Sierra Minera. In ancient times, the area was worked by forty thousand slaves. They would have been Iberians. Some items of clothing and tools have been preserved from those times. Slaves wore sandals, knee pads and two types of cap, close bonnets and those with some material hanging down to cover the shoulders. They were woven from a fibre from the saw palmetto plant, which is common on the hills. The soles of the shoes were not dissimilar to the modern espadrille judging by what remains. The kneepads were woven from esparto grass and attached with cords. They have not been recovered from the Sierra Minera mines, but some were found in Mazarrón.

Many tools are preserved in the archaeological museum in Cartagena, quite a variety of pick like tips to be put on a wooden shaft. The miners also carried woven baskets for minerals and water and a canteen for drinking water. Anything that had fluids in had to be made impermeable with resin or pitch. The weaving of esparto in these tools was superior to that of modern times. I have found modern esparto baskets in one of the mines. The design didn't really change but the quality went down. The La Fortuna mine in Mazarrón where the ancient knee pads were found also yielded a full basket of minerals left by a worker for some unknown reason.

The history of mining is almost as old as the history of man. The Sierra Minera or the Sierra de Cartagena, formerly known as the Sierra de San Ginés, was mined from ancient times.

Thousands of years ago, the Tartessans traded minerals with the Phoenician and Greek visitors to these shores. Nobody knows exactly when it started or who first discovered the techniques involved in mining and extracting the ores. Probably there was some interchange of ideas as well as trade. Beautiful silver coins from the Carthaginian era exist in a museum in Madrid. These were obviously made with silver from the area.

The earliest visible remains of mines date to the Roman period. Early mining was far less complicated than the modern type. The elaborate headframes and machinery that allowed miners to delve ever deeper date from the nineteenth and early twentieth century. The history of mining is closely involved with the history of the discovery of metals. We lightly use phrases like golden age and bronze age to describe times past. In reality, every metal, common or rare, has had its age and has changed the face of history with its accompanying industry.

The Sierra Minera is rich in the ores of many different metals and each has played its part.

Rough dates can be given to the discovery of various metals. The metals of antiquity were gold (6000 BC), Copper (4200 BC), Silver (4000 BC), Lead (3500 BC), Tin (1750 BC), Iron smelted (1500 BC), Mercury (750 BC). All of which can be found in the

Sierra Minera, albeit the gold is in such tiny quantities it has never been of importance in local mining history. Mercury also is not in great supply. There is also the comparatively modern zinc present in great quantities in the Sierra Minera. It is discussed at length in later chapters.

The Visigothic period seems not to have been significant for mines. A small amount of mining continued under the rules of the Mussulmans, mostly metals and alum.

Mines of gold, silver, mercury, copper sulphate, lead, iron and other metals are referred to as appertaining to Cartagena and up to six leagues from it in a document from 1527, in the reign of Carlos I. It is probable that this was fool's gold (pyrite).

Alum continued to be mined and concessions in the El Garbanzal area in the reign of Carlos II refer to silver, lead and gold. Again, this was probably only the fool's variety. In 1769, a tunnel was dug to drain an old well and people thought they had found gold before the authorities closed it down.

Things changed largely in the nineteenth century. The loss of American colonies meant that it was important to source minerals nearer home. The Ley de Minas de Fausto de Elhuyar in 1825 allowed greater freedom. Mining could be started without a special licence from the King. This was a period when small investors could start easily. This was probably one of the least exploitative ages of mining before the big bosses made their workers' lives hell.

The British involved in the war of Independence began to notice Spain's mineral assets. Some of the earliest washeries and foundries were built in 1838 in Portmán: Carpena, Orcelitana, Santa Adelaida, etcetera. British capital was invested there in mines that had been first exploited in Roman times. Spain's industrial revolution was later than that of some other European countries. Consequently, technology from Britain, France and Belgium was involved in Spanish mining of this era.

In 1839, General Requeña put in motion some of the works on Cabeza Rajao. Again, these were using old Roman exploitations. Various small companies were formed to exploit the mines in Roche, Portmán, El Garbanzal, Media Legua, El Gorguel, Sancti Spíritu and others a few kilometres distant from the Sierra Minera such as Algameca and Roldán. 1843 to 1846 was the period of the greatest reopening of the ancient workings. From the 1840s the Parisian Casa de Rothschild began to get involved in Spanish mining. From 1846, there was more working of the superficial carbonates. This was the beginning of the great mining boom. From 1850 until the great closedown in the late 1980s and early 1990s the sulphates were also exploited.

Cabezo Rajao is mainly known for its impressive array of mining architecture rather than its minerals. Mining dates back to before Roman times, but most of what is visible is nineteenth or twentieth century. Photographers love this spot. Its mining wheels turn up endlessly in promotions of La Unión's tourism, although it is actually part of Cartagena. It's not considered a good site for collecting minerals, mainly because outside tips were worked as well as the mines, using up anything of interest.

The mines themselves are mostly flooded and inaccessible.

Cabezo Rajao is a small hill, 190 metres high, a spent volcano complete with crater. It lies opposite the La Esperanza area on the outskirts of La Unión and alongside the N332. It's a short walk up it. There are few plants there. The earth is mainly trodden mining spoils. Rajao means split. This large break in the hill is in part the result of it being a volcanic crater and partly the legacy from ancient mining operations. Slaves were working it in the second century BC. Strabo speaks of forty-thousand slaves in the local mines. The modern mines on the same site go very deep compared with many in the Sierra Minera. Some have up to fifteen levels. A remarkable amount of industrial architecture is preserved. There is talk of turning it into a mining park, but nothing ever gets done. At the moment it is easy to wander up from the highway and see it all gratis. A formal mining park would take years to create and presumably charge admission. Amongst the mines visible are Don Carlos, María Jesus, Trinidad, San Isidoro and Monserrat.

In ancient times, Cabezo Rajao was chiefly mined for its galena containing silver and lead. None is visible on the surface these days. It was necessary to mine deeper and deeper in recent times. Workable veins occurred in areas where volcanic rock met sedimentary rocks. A man told me a friend of his had been given permission to dig there and found raw silver thirty metres down. In modern times, the area was also mined for zinc and alum. Francisco Dorda Lloveras bought Cabezo Rajao in 1852, including the following concessions: Iberia, Virgen de Monserrat, María de los Ángeles and Santa Catalina. In 1877, the mines were leased to Francisco and Fulgencio Martínez Conesa, who modernised the mining and deepened it considerably. At this stage, the first headframes and steam engines were installed. There were further innovations in 1892, by the company, Mancommunidad de Heraderos de los Dorda. In 1924, they were the most important lead mines of the regions.

In 1965, they were bought by Domingo Jiménez Campillo. In 1972, they were bought by Minas de

Cortes, S.A. affiliated to Real Compañía Asturiana de Minas. In 1980, they were closed down permanently.

They were bought by the Sociedad Minera y Metalúrgica de Peñarroya España and some more investigation done but no mining restarted. They were then sold to Portmán Golf with the rest of Peñarroya's assets.

These mines had some of the Sierra's deepest wells. That of Iberia is 434 metres and that of Nuestra Señora de Monserrat is 454. There were many flooding problems in the area. Throw a small rock down one and you should hear a splash rather than a thud.

The fourth winner of the mining lamp in the famous Flamenco festival of the Cante de las Minas, Eleuterio Andréu Martínez, worked in these mines. It's ironic that he was back beneath the surface toiling away next day after his win. In those days, becoming a singer was a way out of mining and the path to a longer healthier life.

There is some footage of this singer on YouTube on a programme on La Unión's music. It is part of the series Rito y Geografía which covers many aspects of Flamenco. He is very old in the film. He was a miner for 31 years and his job involved filling 30 buckets a day, sweating all the while.

It was impossible to combine singing with that, he said. Singing came afterwards alongside massive consumption of wine at the end of the day. Bosses wanted people working rather than singing. Although before the advent of mechanisation there was some singing along with drilling, hammering, or swinging a pick.

Some other forms of Flamenco grew up alongside the blacksmith's trade. The regular swinging of a hammer provided the beat. The unaccompanied martinete is a form of this kind but is not heard often in the Sierra Minera, In the mines, those who carried explosives also sang.

The speeding up of work with noisy machinery was the death knell to the singing of mineras.

A model-maker who is in a wheelchair has made a beautiful model of the Mina Monserrat headframe, several feet high. The original is probably the best-preserved headframe in the area. There are several men who are skilled model-makers. Most of the others opt for tinier versions. You see these together with minerals decorating some of the bars and restaurants round La Unión. Visitors also buy them as souvenirs.

Another small hill close by, Cabezo Agudo, was excavated in 1943. The outlines of rectangular buildings were found, probably those of houses and stores. Ceramic remains and coins place settlement there between the second and first century BC.

The flat land beside Cabezo Rajao, known as Paraje de la Torrecica, also contains a vast number of interesting mining remains. These continue to the outskirts of La Unión, headframes, chimneys, abandoned machines and wheelhouses. Again, this is a wonderful place to photograph. The mineralisation is similar to that of Cabezo Rajao and the techniques of mining and processing were the same. This land is part of La Unión rather than Cartagena. In their day, these were amongst the richest mines of the Sierra. The mines in this area include La Ocasión I and II, San Lorenzo, Revolución, El Tranvía and La Artesiana. Quite a lot of tumbleweed grows in the area, which gives it a Wild West feel. The San Lorenzo mine in this area has the distinction of being the last mine and washery to close down in 1992.

La Artesiana opened in 1878, with two wells and a gallery. It was owned by La Sociedad Los Intransigentes and rented or leased to others. In 1890, it was made deeper and more machines installed for extraction of the ore and pumping out water. In 1897, a new rich vein of ore was discovered 300 metres below the surface. Together with its neighbouring mine, San Lorenzo, it was one of the most profitable mines of the time and one of the greatest producers of blende. Only the best was used, and the rest went into tips alongside. All the galena went into production. There was also pyrite. In 1899, one of the wells was deepened to 418 metres. In 1902, all the workers accepted a reduction in salary, to allow it to be deepened to 500, so that mining there could continue. During the Civil War it was worked by the Unións of the Colectividad Minera CNT-UGT. In the 50s it was bought by Domingo Jiménez Campillo. It was later mined by Minas de Cortes, S.A. until its closure in 1972.

Most of the neighbouring mines had a similar profitable history. La Ocasión was not as deep and had a different kind of differential flotation plant alongside, worked by Enrique Carrión Inglés in 1934. Before this it was a gravimetric flotation plant. Gravimetric flotation produced more dust. The minerals from this mine and from the neighbouring one, Revolución were washed there. Its main well was known as Pozo Carmelo. It passed through a similar history to its neighbours in the Civil War and after. From 1972, the flotation plant no longer worked and the minerals were taken to that in El Arresto in El Gorguel. The mine closed in 1985. In its working days many of the workers were just 14 or 15 years old. Most men worked with a handkerchief tied across their mouths and noses because of the quantities of dust generated. The areas outside had other risks from the steam-driven machinery. Beautiful blende samples were found here in its working days.

The mine En El Tranvía was chiefly worked for the Aguirre family whose name is connected now with the most perfectly preserved and restored palace in Cartagena.

Due to periods of careless working it had greater problems with flooding than its neighbours. It was mainly a zinc mine with some galena. Its last owner was Bernal y Castejón, S.R.C. It was closed in 1970.

There was a Manolita mine in the area of which nothing remains. Even its well was filled up with the spoils from Lo Veremos. It was owned in its working period by Pió Wandosell.

As you get closer to La Unión it is easy to see evidence of subsidence in the mines. A friend who works in the School of Mines in the University of Murcia told me that there is a huge subsidence problem in La Unión. Many parts of the town are undermined.

Apart from Cabezo Rajao there were important developments in other areas of the Sierra Minera in the Nineteenth Century. In 1848, the "manto de los azules", the blanket or mantle of blues, was discovered in a well belonging to Mina Bilbao at a depth of 60 metres. This mine is close to the Rambla de Mendoza on the outskirts of Llano del Beal, along the road that heads towards Portmán.

The manto de los azules is used to describe the top crust of silicates in the area, an unusual formation covering a large part of the Sierra with a thick layer of greenalite, magnetite, silicates, sulphates and carbonates. The greenalite was centred in the areas of the Emilia, San Valentín and Tomasa quarries. In some places it was 80 metres deep. There is also a section of San José and the Gloria quarries with a similar formation. Much has been mined away. Lower levels exposed in the quarries are richer in pyrite and limonite. According to the engineer, Ricardo Guardiola, writing in 1904, in the Gaceta Minera y Comercial de Cartagena, this formation, with its veins and pockets of zinc blende and pyrite, stretched across 500 hectares from the Rambla del Avenque to the Rambla la Boltada, from Sancti Spíritu to the plain of Portmán.

In the middle of the Nineteenth Century there were enough British involved in mining and industry for there to be a British consul in Cartagena. At this stage also, a British Protestant cemetery was founded as part of that in Santa Lucia, on ground that was formerly the military Batería Doctrinal de Brigadas.

In 1849, changes in the laws regarding mining caused a further expansion. Two years later, there were 290 mines in production with 45 foundries: Roma, San Juan Bautista, San Isidro, etcetera. Within years this rose to 650 mines. Most of these were producing either iron or silver. By 1861, lead was another much-mined metal. The Casa de Rothschild had amassed huge quantities of lead in 1860 and continued to be interested in its production even after a fall in its value. They modernised their refinery in Le Havre so that it could process vast amounts at minimum cost. In 1877, they signed a contract with Hilarión Roux of the Sociedad Escombreras Bleyberg. They also had interests with other companies in the other main mining areas of Spain. Soon afterwards the Société Minière et Métallurgique de Peñarroya was created to deal chiefly with lead production. Across the decades that followed it absorbed many smaller companies, including that of Hilarión Roux in 1912. His company also had an interest in zinc, with a refinery in Belgium. By the First World War, Peñarroya was the world leader in lead production.

In 1874, The Carthagena & Herrerías Steam Tramway Company Limited had stations at Alumbres for the blende from La Parreta, in La Esperanza for the mines of Cabeza Rajao, in La Unión for the foundries and mines of El Lazareto and alongside the market for the mines in the Cuesta de las Lajas. Minerals from Portmán were shifted via cable by the same firm.

Initially this was from the Mine of Santa Leocadia with an aerial cable of 1994 metres. Some minerals went out in boats also. In 1882, there were 2770 mines registered, of which 1210 were productive. Minerals going out of the port of Cartagena were iron, galena, silver, lead, calamine and blende. The line was extended to El Descargador for the minerals from Sancti Spíritu and the nearby foundries. Ten years later it was extended as far as Llano del Beal and Los Blancos to take all the minerals from the Cabezo de San Ginés and Ponce, the Barranco de Mendoza and the mines around Llano del Beal.

In later years, much more machinery was used to extract metal from lower levels as higher levels were running low. In this period most of the malacates (a horizontal type of wheel) and the headframes were installed. While wooden headframes are nineteenth century, most others date from the twentieth century.

In 1910, another aerial cable was installed in Portmán for Mina Lucera, property of Miguel Zapata. Iron was loaded on to ships using this. In this period, iron, zinc, lead, copper and silver were the chief metals mined.

The First World War created a greater need for manganese-bearing iron. The mines of Cuesta de las Lajas were important then. By the end of the Great War demand fell and mining suffered a setback. Some veins were running out and outside disasters like flooding and cholera played their part. A large mine, San Isidoro, in Escombreras closed and some workers who could not get jobs locally emigrated to Algeria and France or moved to Catalunya. The factory of the Unión Española de Explosivos in Cartagena gave work to others. Pyrite was useful in the manufacture of explosives. Mines with pyrite around La Unión, such as Cruz Chiquita, Aries, etcetera, increased production for a while. Across the next decade or so the economy improved and with it the demand for metals. At the end of the First World War there was some nationalisation of mining companies. In 1927, there was some government subsidy agreed. After the crisis of 1929, where metal prices were low and the ores were becoming depleted, the Rothschilds left Spanish mining. From 1930, Peñarroya became known by its Spanish name, Sociedad Minero Metalúrgica Zapata-Portmán.

During the Civil War, only iron and lead were mined, and their sale was limited. The foundry at Santa Lucia was under Republican control. At the end of the Civil War various plans for washeries with new flotation techniques were implemented. The first of these was alongside the La Ocasión mine in La Unión. The next was that of the Regente group. The manto de los azules yielded approximately 2.5 per cent lead and zinc, or less in some areas, extracted with a lengthy process of flotation.

Zapata's company bought up many concessions, probably at a very cheap price, at the end of the War. The Sociedad Minero-Metalúrgica Zapata Portmán put the mine and washery of El Concilio into action in 1940, using material from the mines of Sancti Spíritu in flotation tanks. An aerial cable was installed there. Other areas went back into production in La Esperanza, La Parreta, El Llano and Portmán, extracting galena, blende, pyrite, tin and copper.

In 1952, the mines of Segunda Paz and Mendigorría in Llano del Beal restarted and washeries were installed by the Brunita and Lolita mines. The last era was one in which open-pit mining gained importance.

Soon afterwards, the Emilia quarry opened, and exploitation went at a faster pace with the processing of vast amounts of ore in Lavadero Roberto. Túnel José Maestre was cut through the mountains to connect the two. Some parts of the tunnel are lined with concrete, others are rock. For many years it has been flooded, but the water is not deep, so it can be traversed with care. The entrances to quarries are sealed currently, so it's a long return walk.

The Gloria and San Valentín quarries started in 1965. Between 1973 and 1977, the San José and Tomasa quarries were added.

The tonnage from open-pit mining far exceeded that from old-fashioned methods used in the mining operations in Cabeza Rajao, La Esperanza, El Gorguel, Portmán and Llano del Beal. Smaller firms closed down or were bought out. To this period belongs the closure of Minera Navidad, Minera Celdrán and Minas Cartes and its subsidiaries. In 1985, Minerales San Juan closed Mina Ocasión and four years later, San Rafael. Eloy Celdrán tried to adapt to open-pit mining with Brunita but in the end had to sell to SMMPE, who by then owned a very large part of the Sierra Minera. They in turn sold all their assets to Portmán Golf, after the contamination of the bay by Lavadero Roberto was exposed by Greenpeace.

Portmán Golf dabbled in mining for a year or two but essentially it was more of a building firm. Property prices were rising and for a while they were able to get permission to build.

In 1994, the land was requalified and investment in the Sierra Minera was promised. From this period date the golf developments in Atamaría, close to Portmán. These are visible across the protected Calblanque Park.

With the growth of open-pit mining, mining and its waste products increased greatly, changing the appearance of the Sierra Minera forever.

Lavadero Roberto opened in 1951. Within a few years the railway was bringing tons of minerals for processing through the José Maestre tunnel.

By 1992, all mining had closed. San Lorenzo, close to Cabezo Rajao was the last mine to be shut down. So many buildings still retain substantial amounts of machinery, the Sierra Minera has a Sleeping Beauty feel to it as if everyone suddenly clocked off work and might as suddenly start up again. I am surprised that so little was sold off at the time and still remains, quietly rusting and rotting in the open air.

The end of mining brought great civil unrest. At that stage there were only a few hundred miners left but other related industries were also laying off many employees. It was a time when many thousands were suddenly out of work.

There were several general strikes, the chalets belonging to owners of Portmán Golf who had bought the Peñarroya assets and closed them were set on fire. The parliament of the regional Assembly was locked in by the strikers and kidnapped as it were, for a few hours. Afterwards a Molotov cocktail set the Assembly building on fire. There were various injuries though nobody was killed in these riots. Ever since these events there has been a nervousness about the Assembly building. Any demonstrations near it need to be booked with the right paperwork, long before.

If you wish to attend an Assembly meeting you need to give in your residency number the day before and show it on the day. It has far more security than the council meetings at the Palacio Consistorial.

I did attend one meeting there under these conditions with friends who were protesting about the building on Monte Sacro. Ironically, the developers were Portmán Golf.

The politician who put our point of view was Teresa Rosique, an eloquent speaker. I noticed when I read a book on the burning of the Assembly that she was one of those present on the day.

8: *El Descargador*

4: El Descargador

In our wanderings close to Pió Wandossell's foundry in the El Descargador area we made our way along a rambla full of mining waste to a small renovated mine that had been turned into a museum. Las Matildes and the nearby Mina Blanca are rather pretty buildings and have been restored. We went over the Las Matildes museum. A Belgian family came in after us. They had little Spanish but some English and were totally at a loss when the museum guides tried to tell them about the various exhibits. We stepped in and helped out, which is how we became friends with Ana.

Close to Las Matildes and Mina Blanca there are also remains of Mina San Juan Bautista and the oven and chimney of Mina Telémaco. Mina Blanca is also known as Mina San Quintin and has a tower alongside. It's exact age is not known.

Ana was employed at the time by the Fundación Sierra Minera who were staffing both Las Matildes and Huerto Pío. Huerto Pío is a botanical garden where local species are preserved. It was founded in Villa Dolores, which was once one of Pió Wandossell's houses. I spent a couple of mornings there doing workshops. The first was about local plants used for food and gave me lots of ideas.

We ended the morning with a sumptuous picnic full of dishes like nettle tortilla. The second was on making perfumes and household fresheners. The curator cooked up masses of lavender in a copper alembic to produce essential oil. We combined tiny drips of this with alcohol and lemon zest to produce sprays for the house and also wove little fresheners with lavender and ribbon. I kept thinking one of those alembics would make a wonderful still for alcohol.

Huerto Pío is a useful place for exhibiting rural skills. It also has a water wheel and a donkey. These old wheels are known as noria de sangre. The blood (sangre) presumably gives an idea of working the animals to death in the process. The donkey there has a comparatively easy life as he is only harnessed occasionally to give a brief demonstration to visiting schoolchildren. Part of the role of Huerto Pío is preserving rare plants of the area like the Cartagena rock rose. Several plants and trees are under threat of extinction. Industry has played its part in this. Places like Huerto Pío keep these plants and knowledge of them going. Sadly, it is rarely open now apart from the odd school visit.

After meeting Ana Rama in the museum, we went to Venta El Descargador for lunch. She happened to be there with her sister. The place was heaving with custom and the four of us had a job getting served, which meant we got talking. Ana gave us a lift back to Cartagena afterwards and showed us Cala Cortina, a spot we didn't know. I vowed to swim there at a later date. It is now my regular snorkelling area.

The trip to Cala Cortina meant, years later, when we had enough cash to move to Cartagena, we chose a house in Santa Lucia within easy reach of that beach. I once wrote a very short list of those who had an effect on my life, some were friends, some were enemies. Ana was on that list. We have stayed friends since.

On our next visit to Las Matildes we contributed a small collection of rocks of various kinds. We also met another collector who did the same, and rather better ones. Again, he was someone we stayed friendly with. Juan Robles was a painter and produced some interesting sci-fi landscapes incorporating crystals and pyrites he had found. He also had a stall at the mineral fair and outside the Cante de las Minas and was rather more knowledgeable than us. Years later, we would like to update our collection with better samples, but it is hard to do so. Las Matildes is rarely open, and Ana no longer works there. It only opens occasionally for school visits. Being in the El Beal area it is under the governance of Cartagena. Once Mina Agrupa Vicenta opened it seemed to give up totally with Las Matildes. Cartagena seems to be curiously indifferent to promoting anything to do with mining history.

Another mine has been renovated but without a museum attached. Mina Santa Antonieta, off the Portmán road, has been improved and some of the wasteland around it replanted. There are so many mines that need this but too little money to restore them all. The local university polytechnic does some projects to do with replanting, identifying flora and fauna and guiding near there. It is not publicised though, and I only know of it because of one of son's friends being involved in this.

This was not our first trip to Venta El Descargador. We had often found time left to eat there. It was convenient being beside the railway station. Dozens of well-aged Serrano hams hung from the ceiling. The first time we went it was too late in the day for them to cook us a paella, so we settled for a plate of wafer-thin ham and cheese. We warmed ourselves with chicken soup with fine noodles and I drank the house wine.

El Descargador became a point we headed for every time we needed a little comfort eating on a chilly day after walking in the hills. It has a long history as a restaurant. Its first name was Ventorillo Guirao. Probably it was solely a worker's café then. The windows of El Descargador look out to Sancti Spíritu and a tip of mining residues, naturally terraced so you can walk almost to the top of it. It bears a resemblance to the Mexican pyramids in shape. There are also smaller hills with caves before the higher rocks of the Sierra Minera. The walls of the restaurant had a few pictures with a mining theme. One plate on the wall represented a mine in Merthyr Tydfil. It reminded me of a story I had once read of Welsh immigrants moving to this area for the mining. They were supposed to have a vocabulary of their own, but I have never met any of them or known anyone who has. I wish I could. Recently, all this memorabilia vanished with a change of management in the restaurant. I feel sad about this as its position makes it central to many mining excursions and the restaurant was popular therefore with collectors. I have many fond memories. The former owners moved to Tapería Edward in La Unión. Some of the mining memorabilia was on display there. But has gone again as that restaurant has changed hands several times.

In the Nineteenth century, those that came to La Unión with a little money in hand were able to rent small houses in El Garbanzal. Those that didn't started their new life in rough caves. The Cuevas Roma, near El Descargador had been inhabited on and off through the centuries. Does the Roma name refer to the Romans or to the Gitanos who found lodging there in the nineteenth century? You can still access these caves in the hills behind the Sierra Minera station. The small FEVE railway was a mining train in those days, carrying the ore from the mines to the port of Cartagena. The stop was then known as El Descargador. The name is only retained in the restaurant. No one lives in the caves now. The caves are clusters of two or three rooms.

There are flaking plaster walls sometimes retaining a vestige of colour. The floors are mud. These are not posh developed cave houses like the modern ones in Andalusia. These are rough and basic. But they provided shelter in their day. Some have pleasant gardens where you can scrump fruit: figs, prickly pears and pomegranates when they are in season. The figs and prickly pears are worth having, but wild pomegranates tend to become small and hard once people cease to tend them. There are also a number of eatable herbs around, perhaps the remainder of kitchen gardens: rocket, leaf beet, purple mistress (a wild cabbage), mallow and bladder campion which is used in some traditional dishes. There are masses of prickly pears. Recently many of these plants have been affected by a plague. One free food source is dying out. Many tiny birds make nests in the area. They may also have provided a food source.

There is a scattering of individual cave houses of the poorer type across the Sierra Minera also but often only a single one in a hillside. They are easy to mistake for mines until you enter them and find they are only a room or two deep.

There are two other groups of them, one was close to the cemetery in La Unión on land owned by Fundación Vedruna. But some of these have now been destroyed in clearing the land to make a new sports area for kids. The others are in Portmán alongside the contaminated beach. These are the most dangerous-looking cave houses I have ever seen. They seem to be cut into slag heaps rather than solid rock.

El Descargador once had a Roman villa somewhere between the road and the railway. Nothing of the excavation site known as "La Pura" remains.

Ceramics of the period were found there. A short scramble from the road that goes towards Llano del Beal reveals various mining remains, Mina Pozo Cuevas and Mina San Jorge, headframe and calcination ovens.

When the FEVE was a mining train, for many years it ended in the El Descargador area. Queen Isobel II came to the town in 1868 to open it.

To everyone's surprise she insisted on going down the mine, Mina Belleza, whose impressive ruins are still visible on the slopes above and slightly to the right of the pyramid. A steep funicular rail track from there ran ore down the mountainside to the waiting trains.

Nothing but a few spars remains of that railway. It is possible if you are mad enough to walk up the groove where it ran. It is singularly steep. I walk it sometimes. It is excellent training for the calf muscles and a fast way up to Sancti Spíritu.

At the Queen's visit, a mining cart was rigged up with cushions etcetera for her to go to the depths below. She was an unusual woman who had ten children, it is rumoured by different fathers.

She was married to a gay man to cover her promiscuity. Her descendants went on to become kings of Spain. No real problem as the royal descent came through her alone. In age, judging by old photos, she looked remarkably like Terry Jones in drag.

The closest building to El Descargador is the ruin of the foundry run by Pió Wandossell. It is up for sale together with a slag heap alongside.

The number of the vendor is half-deleted which doesn't improve its chances of finding a new owner. We sometimes picnicked there and looked at the remains. A huge grinding machine filled one room. It was probably state of the art at the beginning of the twentieth century. It has now been put into storage to avoid the depredations of thieves and vandals.

Other rooms had a motley assortment of relics: an old poster of a kidnap film, a card of St. Francis, packets of black powder, French chemical jars and various hutches where rabbits were kept in the days after mining activities had ceased. The floor of the first room was covered with rabbit shit. In the middle section of the building a sleeping bag lay next to the skeleton of a cow. I wondered about that. Did the cow or the owner of the sleeping bag die first? And were they in a relationship? Cows are not at all common in this part of Spain. It is far commoner to see small herds of sheep or goats. There is a rambling garden containing trees, an olive, eucalyptus, a fig and carob. Freesias appeared there one year and puffballs another.

We also found a broken vase with my son's birth date and the names of a couple on it. My son found it rather creepy that this date of all others should be recorded there. Years ago, this building could have made a wonderful interpretation centre or an eccentric rambling home with a little work. Every year sees it fall further into decay. It was known as Fundición Pió Wandossell or Fundición Dos Hermanos. It was once a foundry where tons of metal were processed. Pió Wandossell came to the area from the tiny village of Alhama la Seca in Almeria. He was already an experienced foundry man and used these skills to amass a considerable fortune processing lead from the Sierra Minera and also from Mazarrón and acquiring many mines along the way. At the end of the land attached to this foundry there is the small building of the El Cielo mine. It's machinery and headframe have become more and more damaged across the years.

On March 7th, in 1916, the Fábrica de Pío or the Fundición Pió Wandosellwas the scene of a tragedy. Fifteen thousand miners were on strike in the area. It was a time of hunger. Metal values had fallen during the First World War and the price of bread was high. Fifteen thousand miners were on strike, desperate for a small wage increase and weekly payment. They were also asking the bosses to pay for the carborundum stones that fuelled their lamps rather than having to take this cost out of their meagre wages.

It was the time of Carnival. Ironically, on this tragic day, a masked ball took place for those who were rich enough for this kind of entertainment. The foundries were guarded by small groups of police and soldiers. When a group of miners and others saw smoke coming out of Pío's foundry they went to ask those at work to join the general strike. Police and soldiers called for more reinforcements and fired at the crowd. Seven were killed and dozens were injured. Members of Cruz Roja and the Siervas de Jesús who worked in the Hospital Minero rushed in heroically to save those who had been shot. Those who died on the spot, or of their wounds or other complications were Gabriel Gutiérrez Sánchez (37), Francisco Carrillo Paredes (15), Francisco Molero Rubio (20), Herminio Añón Martínez (20), Ginéa Sanz Giménez (27), Valentín Escobar Callejón (46) and Ana María Céspedes Soler (45, a woman). Most of the death certificates from the local hospital were written in rusty red ink that looked like blood.

Pío was a sick old man by then and living in Madrid. He had delegated the running of this foundry and his mine to others. During the strike, workers from Llano del Beal objected to smoke coming out of this foundry. They were gunned down by the militia. Pío, who had been one of the kinder mine-owners, was horrified when the news reached him.

On the centenary of this tragic day, the town hall put on a touching event. The local archivist, Francisco J. Ródenas Rozas, gave a talk with many old photos. This was followed by a long poem, part recited, partly sung by Juan Pinilla Martin. The poem was by Pedro García Valdés and it celebrates in particular the death of the only woman involved. Juan Pinilla, is one of the former winners of the miner's lamp trophy in the Cante de las Minas and he was accompanied by that festival's official guitarist, Antonio Muñoz. It was a very touching evening and I cracked up totally and cried. I expect the males of my family were glad they weren't with me in the circumstances. Juan Pinilla followed up with a fandango that also referred to the seventh of March, this significant date in La Unión's history. A couple of councillors from Cartagena were present also. One was my son's former Physics teacher. Afterwards, various grandchildren of those killed or their rescuers and the very ancient son of one of them were given framed certificates on stage. The Mayor also spoke at length. I still have the brochure, which contains the poem and an illustration by Asensio Sáez.

The strike and its tragedy eventually did something to improve working conditions. Miners got an extra twenty-five centimos a day, paid on a weekly basis and the fuel for their lamps was funded also. A street was named for the day and kept its name for many years until the end of the Civil War. It regained the name again in 1980 and a local fiesta for the date was set up.

I saw a certain irony in the fact that La Unión celebrated this date with such respect, while Cartagena, from whose villages three of the dead above had come, was more interested in celebrating the centenary of the opening of the Gran Hotel, a few days before. The Gran Hotel was built with mining money.

9: *The FEVE*

5: The FEVE

The FEVE, which we used for many of our visits to mines, was partly an industrial train. It is why it is such an effective way of seeing the mining remains. It was built to visit a series of loading points or docks, which had funicular railways and water deposits, etcetera. Most of these have vanished, though the mines and buildings associated with them remain.

It even figures in the mineras, the songs of the mines. Never heard this one sung. It is not particularly poetic, but the words run:

> De Cartagena a Herrerías
> han levantao una pared;
> por la pared va la vía,
> y por la vía va el tren
> y dentro la prenda mía.

The rough translation is this: From Cartagena to Herrerías (a district of La Unión) they have raised a wall. On the wall goes the track. And on the track, goes the train. And inside it is my stuff.

The FEVE still runs but there are other points in the Sierra Minera where a kind of raised wall shows where tracks once ran. These have long gone to the scrap metal dealers.

That is unless the wall refers to the wall put for safety between the main road and the tracks.

Until 1969, the FEVE was a steam train. It was known as El Chicharra. Chicharra is a name for a cicada and refers to the train's slow speed and noise. The original journeys between Cartagena and Los Blancos took one and a half hours. It was so slow in some stretches that passengers could descend for a quick piss and then get back on. Children sometimes hung on behind for a free ride between stations.

These days the FEVE only takes half an hour for the entire stretch between Cartagena and Los Nietos. In the old days, there were problems if the rails were wet with rain or dew so a worker was delegated to put sand on them before the first journeys of the day. The first carriages were wooden.

The Cartagena and Herrerías Steam Tramways Company Limited (British) took over the railway in 1873. (They also became owners of the road nearby in 1900.) In 1882, the line was extended to the El Descargador area. In 1883, a short extension at the Cartagena end connected down to the port in Santa Lucia. A few rails remained for a while. One section ran under the Mompean bridge near the station. It is covered in weeds these days and occasionally gets flooded at which stage frogs move in. Beneath the bridge is considered an area where junkies hang out. This section of the railway used to connect with various workshops and factories in Santa Lucia where silver was separated from the galena. Across the road some other rails ran into the port. These were only recently removed when they were improving the pavements and cycle lane. This rail connected with the docks belonging to Ignacio de Figueroa, Marqués de Villamejor. Within a few years a second branch connected to the dock of the heirs of Valarino, then another to the dock of

Estanislao Rolandi Bienest. Estanislao Rolandi was the proprietor of several mines in Sancti Spíritu: San Sebastián, Recompensa, Olivares and Neptuno. The last rail connecting to the docks was that of Pedreño-Aznar. The old names linger on in a couple of palaces in Cartagena. The Palacio Pedreño is now a bank with an arts centre above and Palacio Aznar has been turned into the church of Antonio Claret.

In 1897, the line was extended from El Descargador to the Los Blancos area. This appears to have been another line owned by a Belgian firm, the Compagnie du Chemin de Fer de la Sierra de Carthagene. Eventually the two lines were connected and consolidated as you see them today. There are stretches these days with just one set of rails and other parts with two. The last stretch to Los Nietos wasn't put in until 1976.

The funicular railway down from Sancti Spíritu to El Descargador was owned by the Compañía de Portmán. It was abandoned in 1894.

There were two branches for loading bays in the Cuevas de Roma area by El Descargador. These were for the pyrite, iron and manganese belonging to Figueroa and Casciario. There were also five loading bays in the Los Blancos area and four in Llano del Beal to deal with minerals from the Mendoza Cargadero. The upper descargador had six loading bays and the lower part, one. There were two more loading areas by the La Unión market station. The old station had three loading bays and La Esperanza, two.

There were also a great many water deposits along the line. The largest was in Santa Lucia. Others were at Los Blancos, Fundación Pura in the El Descargador area, Alumbres and El Abrevadero. Another large water deposit remains in the Lo Campano area not far from the cemetery and next to a marble works. This is huge and covered and looks more like a military building. Few people now know what its use was as it is not even close to the railway.

The mineral-bearing trains were much longer than the current two-coach ones. The engine could pull up to sixteen loaded wagons. The mineral trains and the passenger trains had a different type of brakes. As the mineral business declined, the business passed into the hands of the two ayuntamientos, La Unión and Cartagena, During the Civil War it was hit during an aerial strike in the Villa Paris area a little way out from the main Cartagena station. Several people died and others were wounded.

From 1965, the company became the Ferrocarriles Españoles de Via Estrecha, the FEVE. It was a passenger-only train. In the floods of 1972, sections of the track were covered with mud, especially in the part near the cemetery of La Esperanza. In this year, the new station was built in Cartagena and all the old workshops and the Santa Lucia stretch went completely out of service. Many of these buildings still exist in an abandoned state. A new station was built at El Estrecho, replacing that of Los Blancos, the quarries of which are nearby. In 1976, the final lap to Los Nietos was opened.

I only knew the FEVE from 2002. In the last fourteen years the service has declined in some ways. The trains were renewed and a new station built for the hospital but the stations have less facilities than formally and are more often vandalised. The tickets are available by machine in Cartagena, but nowhere else except on the trains. All the buffets have closed down and the stations look deserted. From time to time, usually just before elections, councils bring out ideas of further laps of the line being built to Los Urrutias, Cabo de Palos or Portmán. So far nothing has been progressed.

The FEVE has another branch in Asturias. They advertise it with pictures at the same time as they are announcing stops at local stations. Perhaps they have pictures of the Sierra Minera running in the one in the Asturias. I would like to think so.

Announcements in Cartagena run in Spanish and English from some central source. Even when a train is not there it is announced as leaving. All very surreal. The guards and drivers are the same regulars that have worked there for years. Since 2012, it has been part of RENFE, the main railway company of Spain.

If you are English, guards tend to assume you are travelling to Los Nietos when they ask your destination. The villages between are all Spanish. Some expats like to leave their cars at Los Nietos then travel in by train to Cartagena. There is a kind of mass delusion amongst expats that there is no parking in Cartagena.

I see English families viewing the mining remains as the train meanders past Cabeza Rajao. Invariably a mansplainer from the family pretends to be knowledgeable and spouts some shit about what was mined there and when. Many of them believe it was gold or copper. I leave them to their delusions.

There is a nice informality about the ticket collectors and drivers on the line. They usually use the second person singular when they ask where you are going. One also begged an orange for the driver off a guy with a box of them, then borrowed my Swiss Army knife to peel it. On another occasion one admired my sunglasses and asked if I wanted to join the local cycling club. It was a kindly thought, but I know I am far too slow for any cycling club I have seen in Spain, especially as I usually end up pushing my bike when hills or mountains are involved.

10: Ermita de San Onofre Monte Miral

6: Llano del Beal

The village of Llano del Beal came about in the mining boom of the nineteenth century though most of the houses there now are moderner. Before then, the population of the area was chiefly near the monastery of San Ginés de la Jara, thanks to the agricultural development of Rincón de San Ginés which brought new settlers. There was also the old village of Beal. In 1859, Beal had 860 inhabitants and in 1887 there were 5632. In 1920, it was divided into the following districts: Beal, Los Blancos, Casas-Cañada, Casas de Emiliano, Casas de Espín, Estrecho de San Ginés, El Llano, Los Nietos Nuevos, Playa de los Nietos and El Sabinar. Some of these areas are hardly populated now and Los Nietos is mainly a summer town with many expats. It is close to dead in winter. It has shallow sea that is safe for young kids to swim in which attracted tourists, or did until it became severely polluted. Little evidence remains of its mining connections. It is an old spot though where occasional Roman remains are found. It was once the site of a Roman fish factory making the highly-prized garum. There were several of these spots on the Sierra Minera near Playa Honda, Los Villares in the San Ginés area, the Island of Escombreras and Las Mateas near Los Nietos. The type made in Southern Spain, was known as garum sociorum. Differing explanations of the sociorum appear in various books. To some writers the word suggests that it involved other fish as well as mackerel. In another book it was claimed that garum production was controlled by a group of knights and these were the "associates". Garum is mentioned by a variety of authors including Martial. I was amongst a group of translators in an anthology of his works produced by J.P. Sullivan. I didn't translate any of those relating to garum. I concentrated on many of the short sex-related epigrams.

Not many people can claim to have tasted garum, but I have. Ten years ago, when I was in a crowd of people demonstrating to save Monte Sacro from the builders, the archaeologist Javier García del Toro turned up with a large Tupperware container of garum he had made. It reminded me of the feeding of the five thousand. A tub of garum and a few packets of Mercadona crackers can go an awfully long way. There were a couple of hundred or more of us in the crowd. I kept going back for seconds and thirds as the garum was very tasty. The nearest things to compare with it are mackerel pate and Gentleman's Relish. I was brought up to believe garum was something rather stinky, but evidently the Romans knew a thing or two. More recently I have found Professor del Toro's thesis on garum on the internet. It seems that it could have an earlier than Roman origin and hark back to the days when the Tartessans were trading with the Phoenicians and Greeks. Garum could also have been part of the tribute paid by the vanquished Carthaginians to the Romans when it was a pricey delicacy, the caviar of ancient times.

Cartagena has exceptionally fine fish. Probably they were even more plentiful in ancient times.

According to the thesis, garum contains both mackerel and tuna. Everything is pounded, including the head of the mackerel and guts. Cartagena has a variety of fish from the Scombridae (tuna and mackerel) family off its coasts. The largest tuna are caught by boat and often sold to Japan as they are very high quality. There are various traditional methods employed also, like the almadraba where the fish are lured into a kind of labyrinth of nets. The nets are only hauled in after a week or two when they are full. Some tuna ends up in the local restaurants. There are also smaller fish from the family like bonito or "melva" which might be no more than a kilo or two. Melva is sometimes used in the local tapas. The smallest form of mackerel is not prized as it is too bony. It is known as "jurel". It is best used for flavouring in soup as it does not have much meat on it. The bigger jurela, (caranx rhoncus) false scad or yellow horse mackerel is delicious. They are fairly frisky to haul in on a line. They tend to feed close to the bottom, so it is almost an accident when you catch one. I have not seen the type of mackerel I used to catch in Hastings close to the shore in the Mediterranean. A few days before I was due to leave for Spain, I had to fill my freezer with mackerel I had caught. They were almost suicidal in their desire to throw themselves ashore. Maybe scomber scombrus became rare thanks to large scale garum production, or maybe the Romans used the other types that are reasonably common these days as well. I do see horse mackerel occasionally amongst fish caught far out at sea. Some are super-sized.

Fish-farming existed in Roman days and there are still some structures visible in the Mar Menor which date back to then. It is not a modern concept. Perhaps some small fish of the tuna family and mackerel were fenced in ready to make the prized garum.

The best garum seems to have started with tuna entrails pickled with salt and allowed to dry out with the sun. It was a process of months.

Perhaps smoking fish took over from earlier methods of salting and sun drying in the fishing industry. The only mackerel pate I have made used smoked mackerel.

The Salinas in various flat areas probably provided the quantities of salt needed in the process. Garum, apart from being an expensive commodity, had medicinal value as did the refuse produced in the making of it. Pliny states that it was good against bites of dogs and crocodiles, ulcers of various kinds. It was high in vitamins and amino acids. What you could call a superfood.

The final three stops of the old mining FEVE train are all in areas of Los Nietos. The last stop abuts a wasteland of saltmarsh, which attracts rare birds. An area of this, known as Lo Poyo, was scheduled for a totally unnecessary development: a golf course, hotel and urbanisation. It has been the subject of bitter rows. The natural part needs to be preserved and the rest is severely contaminated with heavy metals from mining residues. Luckily the EU overrode these building plans at one stage, halting work and insisting on an expensive decontamination. Everything now lies in limbo. I spent a morning down there with other volunteers from ANSE planting glassworts to reclaim some of the damaged land. The contamination is seeping towards the Mar Menor through paths that have been cut, and replanting is the only hope of keeping it from the sea. I only learned of Lo Poyo and its problems when I was involved with a group in local politics. Lo Poyo may in fact get a reprieve due to a corruption case involving the development.

Llano del Beal was the starting point for many of

our walks. Llano means plain. Some have theorised that El Beal bears some relation to the god Baal and older pagan times. Pliny and Polybius mentioned a mine called Baebelo in the days of Hannibal. It produced huge quantities of silver. No-one is a hundred per cent sure where it was but a proximity to Cartagena has been suggested. Maybe the names are related. The El Beal part of the village is across the main road and less interesting. The houses straggle out into the countryside and there is less of a sense of community.

Even a couple of decades after all mining stopped, Llano del Beal was very much a mining village dominated by the slag heaps and ruins of mines on its outskirts. It's a place where you always hear politics discussed. Most of its inhabitants are far left of PSOE. An independent political candidate I know had rather a tough time when he tried to campaign there. During the same election period, I saw two pensioners, one disabled, get into a rather undignified fight in a bar while discussing local candidates. The spirit lives on from the revolutionary strike of 1898 and that in the El Descargador area in 1916. There were further protests at the very end of mining in the area, in 1991, where they played a part. The spirit of protest still lies beneath the surface in this quiet village.

At one time there were mills to extract the water from the mines, but these have disappeared. In 1927, mining activity suffered a crisis and declined. In 1957 techniques changed with open pit mining by the Sociedad Minera y Metalúrgica de Peñarroya, Aggressive quarrying put the local houses at risk. In 1988, Sociedad Portmán Golf S.A who had bought Peñarroya's assets closed down mining in the area. Some other mines went on for another two or three years before the final shutdown. In 1991, a scheme was approved for wiping out Llano del Beal to mine the area and recreate the village elsewhere. The answer to this from the population was: "Queremos vivir y morir en el pueblo donde hemos nacido y solo podrán sacarnos de aquí con los pies por delante." (We want to live and die in the village where we were born, and they can only get us out of here feet first.) So that was the end of that idea. This was the final straw in a period where not only mining, but industry in general was in real trouble. On the 12th of February, there was a confrontation between protesters who tried to shut themselves into the Palacio Consistorial in Cartagena and the police. Eleven were injured.

Llano del Beal is also a place where old customs are observed. You sometimes see partridges in small cages being fattened for roast dinners. They are snared from the surrounding countryside. The first time I posted a photo of one of these birds on Facebook, people begged me to release it. But I didn't feel I could release someone's dinner. Other sporting men there indulge in Colombicultura. This is not pigeon racing in the British way but racing after a female. Large sums of money are bet on the birds, which are all painted with individual patterns of colour. No two birds are alike. It's a bit like the system with clowns' faces. The first time I saw a cloud of these pigeons fly off a nearby hilltop I could hardly believe my eyes.

The belen (crib) in the Casa del Pueblo at Christmas has a cagon unlike the politer versions in Cartagena. The inclusion of a cagon or caganer is a Catalan tradition, which spread to other parts of Spain. The little figure of a shitting man is supposed to be a fertility emblem. The Casa del Pueblo is a gracious building designed by Victor Beltrí. It was from here that the strikers left to speak to the workers in Pío's foundry on the ill-fated 7th of March.

The first few times we walked through Llano

del Beal we passed a restaurant in a private house, totally full of people. Delicious smells were wafting out. We were unable to eat there at that stage, as it was fully booked. El Chupa was an extremely likeable octogenarian who ran this restaurant at weekends. The café in his house had peeling paint, a dog under the table and paper tablecloths but the food was exquisite. It was visited by families from far afield at the weekends. You had to book ahead for one of his usual dishes: caldero, paella, garlic rabbit. The starter was always clams followed by huge meatballs flavoured with cinnamon. Dessert was usually home-made pan de Calatrava, a Murcian speciality which is custard on a cake or crumb base shaped like a loaf and chilled and sliced and served with cream. Restaurants that have translated menus sometimes describe this as bread and butter pudding, but there is little resemblance.

The first time we went, Chupa squeezed us in without the usual reservation as he had some chops going spare. The couple on the next table had come all the way from Lorca. They were having a specially ordered garlic rabbit. The potatoes had been roasted alongside and absorbed all the flavour. They invited us to try a bit. Seeing we liked it, Chupa gave us a portion to take home. This is the sort of generosity you don't see when you eat out in Britain. We paid a few more visits after that. Chupa had a couple of younger male helpers for the cooking and came around the tables telling jokes to his clients. Eventually, health problems forced him to give up. A sad day. He is dead now, but still remembered though his house and restaurant have been demolished to make way for a new building.

Just above Chupa's there is a small park. I noticed a woman there setting a sort of trap for cats and I asked her about it. Her trap consisted of an upside down Mercadona trolley with slight adaptations.

She put some cat food inside and a grid on a string would trap any cat that entered. There were too many cats in Llano del Beal, she explained, so she relocated them to La Unión, four kilometres away. She had adopted some, but it was getting out of hand. Some of the cats would reappear a couple of weeks later. It took them about that long to walk there. She would then begin the process of relocating them yet again. Most of the cats in Llano del Beal and in La Unión are beautiful half-Siamese.

When I first started walking from Llano del Beal there were always huge clouds of dust from lorries on the early stretch of the road to Portmán. They turned off towards a group of headframes, the wooden or metallic structures that cover mining wells and were used for haulage of minerals and miners. There are hundreds of these structures scattered through the Sierra Minera. Some are in better repair than others. These ones, which belonged to the mines, Segunda Paz and Primera Paz, look slightly damaged. They are a wooden structure with wheels. They are in the Rambla de Mendoza. The lorries, which have disappeared now, were relocating part of a heap of contaminated mining spoils. Portmán Golf had given a stretch of land containing these to the council in 2001. The main heap was known as Balsa Jenny, 60 metres high and containing heavy metals. In time it was considered a health risk to Llano del Beal. The gift was likened by some to Snow White's apple. Portmán Golf was eventually awarded the contract for clearing it away, for several million. This was thought irregular so the company Tragsa got the job and then subcontracted to Portmán Golf. The contaminated earth was relocated to the quarry, Corta Blancos. No waterproof lining of the quarry to contain contamination appears to have been done. Some of this earth now covers an area where friends used to find anglesite Gradually many rare and worthwhile minerals are being hidden in this way. Sometimes, but not always, mineral collectors have a last chance to rescue what they can. Balsa Jenny is now in the news again while all the various sources of contamination for the Mar Menor are being considered. Contaminants in the old mining villages have also become an issue again after a high level of heavy metals was found in some children's blood samples. The general level of metals in blood there is not unusually high but the pharmacies report selling a great many inhalers. A new campaigner, José Matías Peñas has come to the fore posting interesting videos re the contamination of areas such as Balsa Jenny. All of this information circulates on Social Media, there are protests and occasional attempts by one or other councillor to ask questions about the contamination and its relation to local health. Yet, little is actually being done to improve matters beyond some fences put around some of the questionable areas.

On our first walk along the road that leads towards Portmán we had a strange encounter with a relic from the past. Sticking up from the mud at the side of the road there was a slanting thing that looked like a bleached, dried stick exactly in the shape of a cat's arm and paw. I pulled on it, thinking it would make a funny souvenir. To my horror I soon realised that it was more mammalian than woody and totally mummified in the alum-laden soil. Recent flooding had exposed this portion of the body, slightly larger than the average cat's arm. We looked at it. Could it perhaps be part of a lynx that had lain buried there for decades? They are now extinct in this part of Spain and rare in other regions. The Iberian lynx is a beautiful creature and only slightly larger than the domestic cat. I took a photo, which came out somewhat blurred, but gave a sufficiently creepy impression of the pale raised arm protruding from the grey mud and sand of the ditch. We had both handled the dead arm and given it a good pull. It was another four kilometres or so before we could get to water in a restaurant. After that experience, I always carried some antibacterial wash. You never know what you are going to touch. Trying to be a correct Mum I handed this to my son when we were on a mineral excursion and picnicking with Spanish friends. They laughed and told us that dirt on sandwiches was good for us... Mostly, I have found it is. The only time I regretted not washing my hands was when I collected minerals containing arsenic in Almeria.

11: El Lirio

Fiona Pitt-kethley

7: Cabezo de Ponce, Peña de Águila and the Campos de Golf

One of the most elegant groups of ruined mining buildings is known as El Lirio and lies in a little-known area between Llano del Beal and Atamaría. Lirio means iris. I am not sure how it got that name. You get there by turning off the road to Portmán from Llano del Beal. There is a pathway to the left close to the top of the hill. It leads you to a wooded area with small old mines and slag heaps. An even smaller and very steep path reconnects with this one at a later stage. If you take the steep route you pass some small abandoned mines and can go up and over the hillside of Cabezo de Ponce with spectacular views en route.

The El Lirio mines have impressive structures but are not much visited as they are not near a road. They can also be reached by taking the old road that winds up from El Estrecho alongside the Los Blancos and Sultana quarries. One of the buildings has the look of an abandoned chateau with reasonably well-preserved ramped passages to the top of it. Presumably some sort of conveyor belt system took minerals up and through these from the mine below. They are impressive and in a reasonable state. You could probably walk up them. A fragment of conveyor belt is still visible in the mine opening nearby. At the front of the building there's a series of shallow curved steps. Sadly, the architects of mining buildings never seem to be recorded, though some of them left some gracious architecture behind that deserves preservation. I think it's a certainty that whoever designed the beautiful ramped passages above arches was inspired by the Rialto bridge in Venice. All this grace for a simple washery for minerals. Its architect certainly deserves a place in the annals of local history. It's a very pleasant area to walk to. In Spring, the route is full of wild rock roses, possible some amongst these are the rare Cartagena rock rose.

Past the El Lirio remains, a large heap of waste sits between tree covered areas. In wet weather a small lake accumulates on top of it. The path and the woods are interesting botanically for the variety of rock roses and for many different coloured specimens of broomrape, which appear for just a few days a year near the mining waste. I have seen more assorted colours of it on this spot than anywhere else, even though it doesn't have any broom to rape. Indeed, I saw no other vegetation close to it, although it is normally a parasitic plant. Perhaps it was harboured on some old roots of plants that had died off near the mining waste. I have also seen yellow broomrape in a different quarry in Agramón in Albacete. Perhaps it is a species that could be grown easily on mining waste.

You can find your way via irregular paths down to the Atamaría area, and the road that leads on to Portmán, passing small long-lost mine entrances en route. Few people make it to El Lirio, this hidden gem of architecture. It requires too much of a hike for the average tourist.

Atamaría and Los Belones can be lumped together as the Campos de Golf. It is largely an expat area. It is the part of the Calblanque Park where permission was given to build. Mixed amongst the chalets and houses are the remains of old mines most of which have barite or calcite as their chief minerals. Most of the population there is English.

I have a dislike of golf courses. They remind me of the set of The Prisoner and seem like a poor imitation of countryside with their lakes and hillocks. It's Nature for control-freaks. Its green grass is always without weeds and has endlessly curving paths plus fences to keep you inside. If you are unlucky enough to stray on to one during a ramble it is remarkably hard to get out again. I was in this position once when I decided to walk from Cabo de Palos to Llano del Beal.

I would presume that most of those who bought in that area like golf and are happy with the outlook. I prefer real unadulterated countryside. From a few angles in the park of Calblanque you see the houses in the distance, stacked up like Lego units.

Most of the mining in this area was for iron and manganese and the mines were worked in the Nineteenth and Twentieth centuries. Most were wound up shortly after the Second World War. The mines are not deep.

The best-known mines in the area are Herculano and Manolita, both for good calcite specimens. I won a piece from Mina Manolita, which was a prize in the raffle at the Mineral Fair. These mines are so close to human habitations now that they are increasingly difficult to visit. Mina Herculano is practically in the basement of a house. Another mine in the area known for its minerals, in this case, barite and fluorite, is Mina Marisol. Its name is not on the 1907 plan drawn by the engineer, Carlos Lanzarote. It is possibly on the concession of La Conciliación. Other nicknames for this mine are La Higuera (the fig tree) or El Submarino because of a submarine-shaped geode found there. It is also known as Mina de las Palomas because of pigeons living in its well. Mina Herculano is right next to Villa 42, one of the earlier houses in the development of the area. Mina Manolita is quite close to the rough track that leads off the road that passes the golf courses and Atamaría. It has a wall that blocks any outside light from filtering inside. Your eyes have to adjust from sunlight to blackness. The entrance has a long narrow ramp with a lot of small stairs in it. It's not the easiest, partly because of its dustiness, the smallness of the steps and the change from light to darkness the minute you step inside. The first flight leads down to a corner with some more steps and a way across to a hole with a spike to hold a rope to help those who go further to negotiate their way down to the next level via a few footholds. In the lower levels there are impressive calcite formations. There's also a heap outside the mine with many small calcite specimens and some barite. If you follow the path a little further and go around to the right, you are in the Rambla de las Nogueras. There are many mines here. I have been in a couple for barite, but the specimens weren't particularly good. Mina Marisol is located here on the right and is the one of these that is most worth exploration.

If you take a right turning from the Llano del Beal to Portmán road instead of going left, it takes you down into a wooded area at the foot of a mountain known as Peña del Águila. There is a chain across the road, and it is marked as a private finca. This deterred me at first as I envisaged it being like a farm with buildings. But the word finca is used very loosely in Spain. After a while I dared to trespass there and found it was an area used occasionally by other walkers and cyclists. This is not so much a spot for good minerals as one to photograph old buildings. There's a huge subsidence problem with almost all the buildings having large ominous cracks. Some are off the paths and you have to clamber among pine trees to find what's left. The only mineral in the area worth collecting is rhodocrosite in micro or small specimens. It is not common, but a little appears in the large slag heap at the foot of several mines, Secretaría, etcetera. The rhodochrosite appears as a pinkish crusting on some of the stones there. From this heap a path zigzags most of the way up the mountain passing several sets of remains. It's a gentle path with several turnings off, some of which lead to mining remains. Most run out amongst trees. At the bottom of the path a large rusty iron tube is visible which passes up a ravine in the direction of the quarries. Most of it is still in place with the odd visible break.

Recently, I heard the owner of Peña de Águila speak at a kind of round table during a conference on replanting the Cartagena cypress. There had been a lack of communication when his land became part of the Calblanque Park overnight. Suddenly he found restrictions on what he could or could not do on the land that had been owned by his family for many years. While the preservation of the countryside seems good in theory, this did not sound like a just situation. In many cases, people are not compensated yet find the uses of their land severely restricted without discussion or warning of what is to come. In the cause of preserving bird life (some eagles nest here) he could not hunt rabbits, for instance. Unless land is bought outright these injustices prevail. There was a sense also that he no longer knew whether he or the council was responsible for cutting firebreaks in this forested area. He was not anti-conservation given that he had turned up for a conference on replanting, but nobody had bothered to tell him before commandeering his land and forcing him to follow a new set of rules.

Peña de Águila is a fraction higher than Sancti Spíritu at 392 metres. The top is less easily accessible.

The last stretch has tiny paths threading through a pine wood. The views from the top are excellent. The last 10 or 15 metres is just rocks where the path runs out. That is where I stopped as I am more of a walker than a climber. It doesn't look particularly difficult though. A graffiti artist had even climbed these rocks to defame the official from La Unión. That graffiti will probably stay unscratched out for many years.

A great many pine trees grow on its slopes but many on the Portmán side fell victim to a fire a few years ago. It has left many blackened stumps of pines behind on the slopes that lead down into Portmán. Small flowers have regenerated at their feet but there is still much to be done there. It is one of the areas where the LIFE Organisation wants to replant the Cartagena cypress.

There are considerable mining remains, such as those of Mina Secretaría. Other mines include Mina Loba's powder house, Mina Santo Tomás with headframe and a machine-house, Mina Candelaria with headframe, Mina Mas Alerta with wooden headframe and Mina San Dionisio with headframe. Every time I am in the area I see further losses to this mining heritage. It's a beautiful but rather sad spot where the evidence of huge cracks in most of the buildings speaks of impermanence. It is obvious that what is here will be the first part of mining history to vanish.

It's a good place to walk in the summer as the trees provide constant shade. In autumn, a few boletus appear there. In Spring there are rock roses in the lower parts of the woods. The whole area has a mysterious light quality. Slight shafts of sunlight penetrating the canopy of trees above give a slightly misty quality to photographs.

Where the zigzag path runs out and is blocked by some large rocks, there is a meeting of three ways. You can continue to the left, onwards and downwards to Portmán passing other mining remains on heavily charred hillside. The path runs out ,but it can be clambered if you don't mind making your way over fallen burnt pines. Eventually you can reach the Rambla de las Colmenas. Another easier road leads across to the quarries. Cantera Tomasa with the remains of Lavadero Roberto II is visible in the distance. It's a silvery path with almost no vegetation. Probably it is only useful on Sundays or holidays when there is little or no invigilation.

12: Opals

Fiona Pitt-kethley

8: Opal

My mother kept a fake opal in her button box. Not sure why she treasured it in the circumstances. It was oval and lilac with a hidden iridescence. I had occasionally seen real opals in rings. They were rare though, as many people talk of bad luck in relation to this stone and won't wear it because of this. Ancient Greek legend connects it with thunder and lightning. It appears where Zeus shed tears of happiness on the mountains after the death of the Titans.

One of my first finds in the Sierra Minera was a small lump of yellow ochre coloured rock with a faint sheen. I only realised this was opal by viewing similar specimens online and in the local museum. The local opal comes in many colours but yellow is the commonest. I could only find yellow and a red that was verging on maroon at first. This form of opal, which is rich in iron, is known as jaspe in Spanish. These colours are derived from the presence of iron in the area. In various spots I later acquired other colours. The prettiest specimens contained a mix of colours. Some are as stripy as tiger's eye. Opal is far less hard than quartz and shatters easily. A large rock can be reduced quickly into many small samples. This is best done with glasses on as the fragments have a tendency to fly up into your face. The small pieces are sharp as arrows and often resemble them. There is definitely potential for forming basic weaponry. I have sometimes found my hands bleeding from handling the sharp fragments. Not too hard to form arrow-heads or knives from such stuff. They would cut for a while but not have such lasting qualities as flint or metal.

One of the highest parts of the Sierra Minera, visible above the pyramid slag heap near El Descargador, is the easiest spot to find opal. This area is known rather mysteriously as Sancti Spíritu. It is 379 metres high. Why was this area called Sancti Spíritu? I have never been able to find out. It is in the nature of high places to be sacred in many religions. Some of the mines between here and Portmán stretch back to Roman days. They are very, very deep. Were there sacred places here in those days? I have heard of no evidence of this. Sadly, Roman mining remains have been buried under tons of sterile waste in this area. A small number of mining remains are still visible in the area: the headframe of Mina Jacinta, the oven of Mina Fortuna, the headframe of Mina Paulina. Sancti Spíritu is mostly just known for its views.

Any opal that is found lower appears to have rolled down the slopes or moved its way via old stream beds. Chemically, opal is closely related to quartz but with more water in its formula. You would not guess it from the difference in their appearances. Under a microscope, they are much more alike. Quartz is one stone that looks pretty much the same either in the hand or greatly magnified. The opaque opal of the Sierra Minera, however, becomes much more transparent when viewed at great magnification. It seems to be a remarkably clean stone. Mud doesn't cling to it. Quartz, on the other hand, often emerges in a thoroughly soiled condition.

Mostly the opal occurs on its own. Occasionally there are veins of quartz, pyrite or galena or tiny inclusions of black glittering goethite. One large block of yellow opal I split included a tiny geode filled with red powder, probably red ochre. The dust of the opal where I have broken it also looks like a fine pigment. One day I will experiment with using it as a basis for paint. Sancti Spíritu is the highest part of the Sierra Minera. There are eight windmill turbo-generators running across the skyline. It is one of the weirdest places I have ever visited. The windmills are visible from many miles away. Close up, they tower and cast odd shadows across the land. You can hear an Aeolian whistling through their blades and the sound of motors turning. One has an unoiled squeak. They are slightly unnerving. The first time I reached them I had to steel myself to walk along those paths. I have visited them on relatively still sunny days. I would hate to be there when there was a storm. The path winds to and fro over several kilometres between the windmills. By the sixth there is a series of satellite masts. Sometimes the path descends and loops round a hillock on the way to the next. On one side you see La Unión below. To the other there are views to poor ruined Portmán and down into vast quarries. By the time you reach the third windmill the road is paved with opal. On my first visit I had only found the golden kind. On later trips all sorts of other colours emerged cream, brown, red, purple, grey, green, blue and black The opal is low quality, nothing like the expensive Australian sort. You recognise it chiefly by its sheen.

Occasionally, you see eagles flying above the windmills. We once discovered a decapitated body of one in a quarry below. The turbine above seemed damaged. It was no longer turning and a sooty mark appeared below the blade.

One day I met a party of walkers from La Manga that made it to the first windmill. Usually though, this area is entirely deserted. It's not green and verdant but it has a kind of spooky sci-fi beauty that grows on you as you get to know it. If you turn away from the windmills you can visit other parts of Sancti Spíritu. The plateau that covers the top of the Sierra Minera contains many valleys and hidden places. It is easy to get lost there. The windmills are known as the eight kings of the air. They provide a point to orientate yourself.

Occasionally, on public holidays, quad bikers race illegally in the quarries. Their heavy wheels grind the paths into tracks of silver. One day, I came across a magical path that seemed to be paved in light blue. I dug out chunks and there was a mix of rough opal and chalcedony. It took me years to find this spot again. This is probably the only part of the Sierra Minera that resembles the romantic phrase "manto de los azules".

One question I have never been able to answer is who makes the paths? The council certainly maintains a section of the road between La Unión and Portmán known as Ruta 33. The first half an hour of this walk uphill is a tourist attraction. There are wooden benches to sit down on occasionally. Gardeners plant and prune the trees on the roads steep edge. The road winds up to an area where there is a small chapel in the rock. Mina Agrupa Vicenta has been restored nearby and can be visited. It is up a flight of steps. In the year while they were building these and restoring the mine inside, I remember seeing a heavy digger left on the edge of the road, almost teetering off the edge. Cassiterite and pyrite were mined in this area. The old vein where tin ores were visible is now covered in concrete alongside the restored mine. There are a few small caves you can crawl into as well, but the minerals are crumbly and hardly worthwhile. Pigeons nest in this complex of caves and mines. These are the fattest most contented pigeons of the Sierra Minera thanks to the figs. In the Middle Ages people ate a small bird called a beccafico, which had a similar diet. Probably these pigeons are exceptionally tasty.

The road and its surroundings are well-landscaped here. There are lavatories, seats and bins for your picnic wrappers. A Spanish flag flies here, which can be seen clearly from below. Many people make the walk to this point, on Sunday afternoons. An old powder house has been converted into a little chapel you can visit and sign the visitors' book. Beyond that the road winds further up the mountain but fewer walkers go further. Eventually it reaches a division of the ways. From here you can find the tiny paths to Sancti Spíritu if that is your aim or descend to Portmán. It's a meeting of three ways. These were sacred to Hecate in the old days. The third path loses itself amongst other peaks with poorer pyrite mines and pine clad slopes.

13: Pyrite

9: Pyrite

The top level of the Sierra Minera is silicates. The layer below contains pyrite. Most of the pyrites sold in mineral fairs come from Navajún in the La Rioja province where they have a beautiful geometric form. I have visited this area and have many samples at home. In the Sierra Minera, pyrite shows mainly as a metallic sparkle though in some areas above La Unión tiny cubes are found. It was these that first attracted my son in the museum.

Pyrite was traditionally mined with explosives rather than picks. Hit a rock with a hammer and you will see why. It needs more than human force to get it out.

Pyrite was at the most poverty-stricken end of mining. You need a lot to make iron and iron is a cheap metal. Children never worked the pyrite mines because of the danger with explosives.

Pyrite can also be used to make fire using the sparks by hitting it close to dried-up materials like desiccated mushrooms. I have seen this demonstrated in the Mineral Fair. It generates a strong sulphur smell.

The restored mine, Mina Agrupa Vicenta, was a pyrite mine. Just a dozen or so men worked it in its heyday. It was originally Mina Agrupa Vicenta and was owned by the Sociedad San Hilarión. Hilarión Roux, Marqués of Escombreras was one of the owners of this company, hence the name. It then became part of his larger company, Companie Francaise des Mines et Usines d'Escombrera. It supplied material to the explosives factory in Cartagena. During the Civil War it was part of a collective. Afterwards, it was part of Minera Celdrán, S.A. This company installed a flotation plant, the remains of which have been restored for visitors. The mine stopped being worked in 1960 and became part of the assets of Peñarroya and afterwards, Portmán Golf. In 2005, it was ceded to the ayuntamiento of La Unión. The mine was also used in a documentary about the Sierra Minera. I watched this in MURAM, the modern art museum in Cartagena. Actors were trying to use pickaxes unsuccessfully, assuming they did not have the strength of past men, not realising that explosives were needed. By this stage, the mine was owned by the town hall.

For guided visits air freshener is used to mask the sulphur smell that comes off pyrite. In the good old mining days this smell sometimes gave the men the shits. An ex-miner's wife told me her husband always had a problem while working in mines of the area.

Mina Agrupa Vicenta has gone up in the world with a ten-euro charge for visiting. The organised tours take an hour or so. Our friend, Ana, was one of the guides for a while. Occasional Flamenco concerts are given there because of the good acoustics. A metal-lined ventilation shaft allows water to drip in from the hillside above as well as that which seeps through the rocks. The small lake near the concert space is entirely acidic.

If you turn right instead of making your way up to the chapel and the ruins of the Pablo y Virginia mine on Ruta 33 beside Mina Agrupa Vicenta, you pass many areas with pyrite. Pablo y Virginia is named after Jacques-Henri Bernardin de Saint-Pierre's novel, Paul et Virginie. Every area had its hidden mines and dried-up paths of streams. We found several tunnels with sulphur encrustations and ground slate and powdered pyrite in abundance. The lowest was flooded. A tiny delta in the silt was draining some of the water away into a dried-up riverbed. The water travelled for a short way before disappearing, sinking into the soil. It was rich in sulphur and iron, coating the nearby rocks and dry thistle stalks which had fallen into the water. We decided to give these channels a helping hand and chipped away at them. Soon the water was draining faster. At first the sulphur turned it completely yellow. Wafer thin flakes of iron floated on the stream. We saw an ant riding one like a surfboard. After a while we were able to wade into the start of the tunnel, stepping from stone to stone. The mud between was treacherous. In the distance we could see a drier part that we hoped to make our way to.

The riverbed outside had the remains of old walls with square holes for water to flow through. The countryside must have been prettier when the water flowed more freely. Tiny pools in the old watercourse made miniature cascades like something from a Japanese garden. A few large stones had tiny quartz crystals on them, stained red by the iron and with flecks of pyrite that glittered in the sun. They were too large to take away, but we enjoyed looking at them with a loupe. This is the mine known as El Tesoro de Carolina.

We considered this a trial run for trying to drain the more famous José Maestre Tunnel at some stage. Below the tunnel, the dried-up stream bed passes a hillside with a curious formation of stones on it. They zigzag across the grass. It is described as a serpentine chimney which once belonged to the mine called Trinidad. It is clearly visible on Google Earth.

Serpentine chimneys were flues that zigzagged along the ground as a way of limiting the contaminants reaching the air. Some of the muck gets left in the loops. There are several of them in the Sierra Minera although this seems to be the most perfectly formed.

I returned occasionally in the next week or two and found the tunnel had less water although it had not completely dried out. Presumably more water was filtering down through the hills. The streambed outside became steep after the tunnel but it was just about climbable. It culminated in an area with a long tunnel mine and rocky walls alongside. We climbed one of these into a space with a lagoon of mud. The tunnel was easy to enter being higher than many but petered out after a hundred metres or so. Again, it was filled with the sulphur and crushed pyrite of the area.

There was a large glob of black jelly outside this mine, which spooked my son out. I assumed, prosaically, that someone had cleaned muck off their hands with Swarfega. On the other hand, it was not unlike pictures I have seen since of mysterious jellies that turned up in the vicinity of meteorites.

One of the mines of this area, Mina Inglesa, was the site of a curious find in 1842, which is mentioned in the newspaper El Espectador. Roman graves with mummies of gigantic proportions and other curiosities were found there. This is an odd tale to which I can find no references elsewhere.

The area is known as Cuesta de las Lajas. Lajas are large slabs or blocks of stone. As I had climbed up the streambed I encountered many huge slabs of slate, some turned on their side by the force of the water that once ran by here. It's a fun area to clamber around if you want something different to a normal walk.

The Cuesta de las Lajas, between La Unión and Portmán, is one of the areas with the greatest concentrations of mining. It is officially part of La Unión, though it is hard to work out exactly where the boundaries are. The map of this part is hard to read because of the concentration of concessions scattered across it. The stony slab-like structure of this area is certainly obvious in parts of it. A smaller section of it, the Collado (hill) de las Lajas contained what were known as "minas de pinturas". These were small mines owned by a Señor Ortega who produced colours for artists from them. There is an abundance of red, yellow and brown ochre colours here. The slag heaps from the foundries provided blacks and greys and there was the green of malachite in some of the quarries nearby. Probably only blue is lacking.

Pyrite exists also in tiny mines accessible through the pine woods. These are used by dog-walkers and mushroom collectors in autumn. I have picked blewits and boletus there. In these woods there is a pleasant area known as El Chorrillo where a series of works have been executed around a spring: stairs, canals, a tiny chapel. They were all the work of one man, Antonio Pagan Lorenzo, who was the uncle of Rogelio Mouzo Pagán who wrote fascinating blogs about La Unión. He executed these considerable works with scrap stone and bricks, only buying the cement. It was done across forty years. The land was in fact owned by Peñarroya but has always been used by people from the town. Decades down the line Peñarroya objected to his alterations and he had to stop. His works have fallen into disrepair since he died, but the town does occasional clean-up operations with volunteers. Sadly, the spring dried up and there is much graffiti and rubbish. The spring started again recently during lockdown.

To the right of this wood there are pyrite mines. A model-maker in the town uses his blacksmithing skills to create replicas of the mining headframes and trucks. They are beautifully detailed. If my cats were less frisky I'd buy one or two as ornaments. He uses tiny pyrites from these mines in the models of trucks. He is one of several model-makers we have met.

Pyrite is omnipresent in the quarries where the top layer has been dug away with open-pit mining. Corta Brunita is the prettiest of these because of its lake. On a still day there are perfect reflections and it is hard to tell where land ends and water begins. Occasionally here people have found blue vivianite, but not in the last ten years.

Pyrite, known as fool's gold, has always been a source of attraction to those that don't know much about minerals. In today's new age interpretation of stones it is supposed to be a money-bringer. At Spanish mineral fairs stall-owners often put "piritas de suerte" on a bowl of them and punters buy them for luck (suerte). I have to confess I have done this myself to sell a few more though I am somewhat sceptical having carried one for years which bore little fruit. One year, some little boys came up to the stall I was keeping outside the Cante de las Minas. They said I must be a fool as I had fool's gold (oro de tontos). I thought quickly and told them it was just for my clients. For a couple of years, we kept a free stall, solely for swaps, at the big mineral fair. In many cases the swaps we got were not great, so we gave this up. One year though, a boy insisted on swapping a little phial containing a particle of real gold for some fool's gold. These little bottles came with mineral magazines as a gift at one time. That's the sort of swap I like.

13: Galena

10: Galena

Galena has exactly the same name in Spanish. It is rich in silver and lead and has been mined since ancient times. It was simple to process as the lead melted off at a much lower temperature than the silver. My first encounter with this mineral was outside the village of Llano del Beal. A dried-up riverbed was being excavated and lined with blocks of stone in case of flash floods. It was one of those ostentatious, and probably unnecessary, works that transfer substantial amounts of money from the local town hall to whichever contractor is lucky enough to get the job. Most of this kind of work has dried up during La Crisis, but a few years ago it was commonplace. Lorries were toiling along the road with clouds of dust blowing from their loads. It made walking this route difficult. The road outside Llano del Beal leads up and over the hills of the Sierra Minera into the Calblanque Park, a protected zone. It is a beautiful walk once you get past the initial dusty stretch. At a junction of roads there was a huge block of grey-green stone with gouts of a silvery mineral embedded. I hammered off a few chunks and took them home as did other collectors in the district. The stone soon vanished. The only evidence left was a grey green silvery powder embedded in the newly asphalted road. It had become part of the landscape. The silver eventually disappeared completely on the wheels of passing cars. One or two people claimed it was the last of the galena. This was of course not true. But it was the last that was easily and freely collectable without trespassing in quarries or delving deep in mines.

I put some chunks of this stone amongst a collection of rocks I was selling or swapping at the monthly mineral fair at San Vicente del Raspeig just outside Alicante. The first guy who looked at it said confidently that it came from San Valentín. San Valentín is a quarry in the heart of the Sierra Minera, which is a kind of Mecca for mineral collectors. The usual road taken to it winds up from Portmán and carries extreme prohibitions and threats of denuncias. The phrasing of these is far worse than the usual trespass warnings. Mineral collectors from other parts often put their photos on the net standing beside these as an act of bravado. In reality, other roads lead to San Valentín too, but these are less well-known. The owners of the quarry don't seem to want visitors although it is hardly used these days. I eventually managed to visit it more or less legally with a member of our mineral group. He had friends or relatives on the gate and in the Guardia. He is a doctor who collects as a hobby. He had spent so much time there he was quite blasé about its joys. He went off on his own and managed to find the most fantastic specimen of crystalline galena I have ever seen, better than any in museums I have visited. It was beneath one of the windmills of Sancti Spíritu. He had climbed up from the quarry below seeing a tell-tale glitter in the quarry sides.

The galena from Llano del Beal was in blocks of greenalite. In the quarry, I found streaks of it in green opal. San Valentín also has a huge array of other minerals.

Some of the most beautiful galena I have seen is in a very small space in a mine that is not too far away from the one I called The Chicken Mine. This is known as Mina Rómulo and is popular with collectors. A ramp bears downwards past a ventilation shaft full of foul air. You can smell the lead distinctly there. Further down there are tiny spaces you can crawl into. I had been told that probably only my son could manage it, which I took as a sort of challenge.

My son should have known I could do it, as he and several children from La Unión lost a game of Twister they played with me. Taking up Hatha Yoga at 10 has left me with some permanent flexibility. I nearly got stuck because I forgot to remove my backpack while crawling into a hole. But the sight was worth it, huge quartz geodes mixed with galena. When I see things as beautiful as that, I don't want to attempt to remove them. They are a sight for posterity.

Galena existed in many other parts of the Sierra Minera and was widely mined by the Romans who of course had plenty of slaves to do the dirty work. Romans never did their own mining. Raw natural silver also existed although that is much rarer now. The Galena of the mines between Cabo de Palos and Cala Reona was supposed to have a high silver content and were mined by the Romans. There are still mining remains near Cala Reona but most of those near Cabo de Palos have disappeared under modern developments. Quite a few of the mines in Portmán were also started by the Romans, such as Mina Segundo Ferrocarril. This is a flooded mine which is full of mud on the lower level these days. It is known chiefly for having flattened crystals of galena.

Galena was also known as "alcohol de vidriero", alcohol of the glassmaker. And so, galena mines became known as "minas de alcohol". It was used as a hardener to form lead crystal similar to that of Bohemia in the glassworks of Santa Lucia. These days, all that is left of the glass industry is the glass museum, which is used by several artisans who continue the old traditions, and also demonstrate their skills for schools and tourists.

Near where I live in Santa Lucia, there's the remains of an old lead foundry where the galena was processed until thirty years ago. It was once the San Ignacio foundry. It was owned, in its latter days, by the firm that polluted Portmán then went into liquidation after selling all its assets to Portmán Golf. The old deserted buildings are still known as Peñarroya locally. Unemployed men dig there in search of scrap metal. It is rumoured that a lot of lead was buried there when it closed down. And kids sometimes play amongst the ruins. On Sundays, a group of adults played war games, firing plastic bullets at each other, using the ruined buildings as cover. It has now been bought by the container port opposite its walls.

Peñarroya is sited at the foot of the San Julián mountain. A German who is now a professor at the local university was the manager there at one time and he told me he had a large bin full of chunks of galena in his office that he gave to visitors as souvenirs. He talked of the closedown of the mining operations as being triggered by a nearby village's unwillingness to be destroyed when his firm needed to bore through the mountain. I realise now he was talking of Llano del Beal.

The lead foundry outlasted most of the rest of the mining interests by a year or two. Portmán Golf was mainly a construction firm and had other plans for the Sierra Minera.

Washing Amethysts in the Bidet

14: Escombreras Industry

Fiona Pitt-kethley

11: Escombreras

Beyond the San Julián mountain is Escombreras. It has just five inhabitants now. They must have some strange immunity to the strong smell of petrol, the humming of pylons and the huge amount of light there during the night. Escombreras is a frightening but interesting place. It's the most industrial area of the region. It has six thousand workers. If World War Three broke out it would be a sitting duck, an obvious target. At its centre is a huge oil refinery, which is within a stone's throw of huge crackling pylons, factories for Eco fuel, hydrochloric and sulphuric acid, fertilizers and explosives. Another refinery of a specialized kind has been built with a huge investment from Korea. Is it arrogance in the management of Cartagena that placed all these industries so close together? Or maybe they are just unwanted elsewhere. Somebody has to have them.

    Several decades ago, Cartagena itself was filled with industry. The air was so foul at night when the factories opened that there was a curfew for the non-working inhabitants. My son heard this from teachers who were old enough to remember the bad old days. What would happen if the oil refinery went up? A fire there in 1969 burned for a week despite the constant attention of local firemen. I have seen the film footage and it is horrific. There have been several narrow scrapes since, when forest fires came close to the area.

Escombreras is a very ancient place and retains some charm even in its tortured state. The island that carried the name earlier in time is separated by a narrow channel from all this industry. It was first known as the Island of Hercules under Greek occupation. The name Escombreras came later, because of the abundance of mackerel (Scombri) in the area. There are still temple and Roman fish factory remains on the island which I would love to see but I cannot think of a way of visiting them from the land. The island is home to a rare seabird, the Audouin gull and still houses a unique plant, the Escombreras chamomile. It used to be common throughout the mainland part of Escombreras, but the quarrying has annihilated every single plant. There are plans to replant it and some has been done in other coastal areas.

The flat coastal section of Escombreras is fenced off and is part of a huge container port. No chance therefore of snorkeling out from there. There are many no-go areas in Escombreras. There are still people alive who carry a sadness for what was lost. There are pages on the net where they remember the good old days. There were many mines there and you can still see some deserted buildings, but many are hard to access because of the security issues.

The modern mining in this area was started in the 1830s, by Hilarión Roux, a Frenchman. In 1874, he built a school and church in the area and is therefore considered the town's founder. There was a cholera epidemic in 1885 which was dispersed by measures of safety and hygiene with 200 people isolated on the Islote de Escombreras His lead foundry passed to Peñarroya in 1912.

Escombreras was a fishing village with orange groves nearby before it turned industrial. Tuna was caught there and there was also a variety of shrimps in the shallows. Some people still climb down the rocks to fish. It's less possible to fish by boat there now with oil tankers and heavy cargoes leaving for all quarters of the world. Close to the end of the Civil War, Escombreras was the site of a famous shipwreck. The Castillo Olite, filled with Franco's troops, was hit by a canon shot from the Batería de la Parajola on the Roldán mountain. The shot pierced it at a spot, which has the curious name of santabárbara in Spanish. Santa Bárbara is the saint of miners and those who use explosives. There is even an organization in La Unión called Los Caballeros de Santa Bárbara. They are all men, no female knights are allowed. They all turn up at the annual mass of Santa Bárbara. Her image is carried, and firecrackers are let off.

On a ship, the santabárbara is the area below deck where explosives are kept. The Castillo Olite rates as the Spanish shipwreck with most loss of life. Of the 2112 men on board, 1476 died, 342 were wounded and others were taken prisoner and moved to Fuente Alamo. The fishers of Escombreras played a part in the rescue of the survivors. According to a website written by people from the area many had lost everything, so they dressed them in their clothes. When these ran out, they gave them their wives' clothes. The woman who kept the lighthouse was also helpful in rescuing the sailors. Her name was María del Carmen Hevia de Saavedra, a beautiful twenty-five-year-old. She became a Francoist heroine and was celebrated afterwards. Judging by a book I read on the shipwreck though there may have been a darker side to the rescues as some looting went on. The wounded men were kept in the small local church and many died. Some would have had a chance if they had been shifted to Cartagena hospitals, but they were there for many hours before being moved further afield to Murcia and other places. This added to the huge loss of life. In the months that followed, corpses galore were washed up and buried in the now abandoned cemetery. Local people ceased to eat the fish knowing only too well what they were feeding on. Shortly after the tragedy, a large stone cross was built and there were yearly commemorations. As the area became more and more industrial the site was moved. In 1956, a large metallic cross was put up instead. Eventually this was demolished and stored in pieces on a building site. The shipwreck was 24 metres or so under water. In 2003, the ship was dynamited to break it apart and large chunks of it were brought up in sections to be sold for scrap. This released many bones into the water from the more than a thousand bodies that were trapped inside. A few small objects were rescued for museum displays.

So little respect was paid to the dead both there and in the local cemetery, which is untidy, abandoned and bears no record of the dozens of bodies buried. The little church also is an empty ruin with no record of its temporary use to shelter the wounded. There is an aura of sadness about the area. The whole story shows a lack of respect. Is it the fact that they were Franco's soldiers? Or is it that industry and the money from it are thought more important than history and commemorating the dead? The survivors of the shipwreck came back occasionally to Fuente Alamo where they had been imprisoned and had better relations with the locals. There seem to have been bad memories of Escombreras, on the whole. I have a DVD with some interviews with these survivors who were in their eighties at the time they were filmed. They are all dead now. It's a fascinating record. Brave men even if they were on the wrong side.

The writer, Arturo Perez Reverte was born in Escombreras and walked the hills as a kid and dived for amphorae off the beach. The amphorae are all now in ARQUA, Spain's first underwater archaeology museum. The different shapes of amphorae that came from the sea here testify to the trade of Phoenicians, Greeks, Carthaginians and Romans. Amphorae were used to trade wine, oil and garum.

For a while Escombreras was filled with workers and a town made for them called Poblado Repsol. You can still visit the ruins of the church and the outlines of a park and school area. With expansions of the factories it was depopulated. Officially only five people live there now. Perhaps they were too obstinate to move. I feel a certain sadness the times I have clambered on the hillside to examine these remains. The church has been gutted, but pigeons in its roof give the sound of a ghostly congregation. The cemetery nearby has very few graves left and is a wilderness. These remains, the military batteries and a few ruined cottages are all that is left from older times. Everything else was annihilated and covered with modern industry. There were 3156 inhabitants in 1970. This fell to 297 in 1996 due to crises in the petro-chemical industry and contamination.

On the outskirts of Escombreras, in a kind of no-man's land tucked between it and El Gorguel, lies the town's waste tip until they find another spot to desecrate. It smells pretty ripe. It is close to the impressive remains of the Reina Regente mine washery and Mina Camarón. Between this and the most heavily industrialised area there are fields where some orange trees and carobs still flourish. Some of the old mining installations and the ground about them have been turned into "cazas de coto". These are private hunting areas. Not sure that these should exist so near the refineries as with firearms there comes a risk of fire. Most of the area underfoot is covered in yellow-flowered oxalis plants whose leaves make the earth look very green and are probably safer, with their dampness than the matorral in other areas.

Rogelio Mouzo Pagán wrote an interesting blog on Mina San Rafael with pictures from the days when it was still a working mine. I always feel slightly envious of those who could see mines working and well-lit with an electrical system. You see so much less by the light of a feeble torch or several. Mina San Rafael (sometimes known as Segundo San Rafael) was worked on two levels. Galena and blende were the principal ores mined there. In 1909, it was bought by a German company, Orcharson and Enthoven, who owned other mines in the Sierra. In 1930, this company was acquired by Peñarroya. They never mined San Rafael but leased it out. In 1949, it was registered in the name of the Sociedad Anónimo Civil La Alternativa. In 1972, it was leased to Andrés Moreno García. In 1975, the concession was sold to D. Manuel Rodríguez Esparza. In 1976, it passed to its last owner, Industrias y Minerales San Juan, S.A. whose boss was Don Juan Conesa. It continued until it was flooded in 1989. Metal prices were low at the time, so it seemed better to close it totally.

There are still good mining remains with some passable micro minerals of quartz, calcite, siderite, blende, pyrite and galena amongst the slag by the roadside. Some have also been found of green and blue barite, marcasite and stibnite. I approached Mina San Rafael after rambling through the woods between the Santa Antonieta mine and Escombreras. I eventually reached a good service road that leads up to the mine and winds on to other remains in the hills. When I walked down this road it was gated near the bottom and I had to scrape round the gate. On the other side I saw a host of prohibitions and warnings of danger about changes of level, etcetera. In reality, the road looked a good deal safer than most I have traversed in the hills. Another mine of this area where El Gorguel touches Escombreras is Mina La Verdad de Un Artista. also known as La Verdad de Los Artistas. It is in a valley with the headframe set on a sort of hillock of mining remains. These can be seen from several vantage points on nearby hills. It's close to the part of Escombreras that looks fairly rural apart from the mining remains on its outskirts. There are still some orange trees in the fields nearby. The area gives an idea of how Escombreras once looked before the industry moved there. The collectable minerals appear to come from a gash in the hillside above the mining remains. It was initially owned by Mineros de Verdad. Few people venture there because of the difficulties of access and the impossibility of parking a car nearby. The security firm that patrols the area is very vigilant.

It is flooded at the bottom, but some people have abseiled down it. It was an iron mine but the only collectable minerals nearby are those of copper. It is the only mine of the Sierra which produces a decent size of azurite crystal up to a centimetre long. Their appearance is much more crystalline than any others in the Sierra Minera. There is a lot of inferior quality malachite also and tiny beads of the lesser-known adamite on some of that. One of the reasons security guards are not happy if they see you in this area is because the old mining road that runs up to the San Rafael mine continues to a small quarry and runs out by gates and fencing beneath what appear to be more oil containers judging by Google Earth. It is not an area you can get close to physically.

15: Stones from the temple of Júpiter Stator on Monte San Julian

12: Lo Campano and San Julián

Halfway to Escombreras, there's the beautiful beach of Cala Cortina where I snorkel. This is home to a mine that almost nobody visits. Judging by the maps it is on the spot where Mina San Águstin was, but it is in fact an underground quarry. There is an entrance behind the car park above the Mares Bravas restaurant. You have to scramble across wasteland overgrown with tumbleweed and other shrubs to find it. I took a picnic in there with my son as a short walk takes you to a cathedral-like space which has some natural light from a shaft leading out to the mountainside. It has some very clear gypsum like a film of glass and alabaster and a little selenite. This level also had an extensive railway. It is possible to clamber down to a lower level which is also spacious and rather like the first.

Sadly, the rails have now been taken for scrap together with a small engine and mining cart. It must have been quite an operation getting them out. So much is disappearing.

In the land outside the mine there is an old water deposit that at first sight looks like a military building. There are also various channels where some old water system was in place. It is all hard to access due to the landslips of rocks and copious weeds.

On the hillside above, I found some specularite, a rare mineral that is the same formula as hematite but which leaves silvery dust on your hands. It showed evidence of iron in the area as do the large blocks full of chalcopyrite that are inset in the rough piers between the three coves that make up Cala Cortina.

Some years ago, a murderer hung out in the underground quarry or mine. Rather glad he was not there at the time I first visited it. A confessional note from him, or a forger, revealed his whereabouts to the police. The murderer is now in gaol. When I went back there more recently I found a string leading through the tunnels to a space where there were some old clothes and half-used red candles of the types left at shrines. Perhaps this was where he hung out, or perhaps they belonged to some later squatter.

The beautiful San Julián mountain which runs down to Cala Cortina on one side and Escombreras on another, contains many mining remains. I found some evidence of copper behind the old deserted Peñarroya buildings. Crushed into the ground near the cave mouth of a small mine, various chunks of rock showed the green of malachite. It is the worst quality malachite I have seen, so altered that there was nothing crystalline about it even under a microscope.

A little higher up the hill, there's a pond with very vocal bullfrogs. It appears to be the remains of a very small quarry. From here a path leads up to the chimney of the old lead works. Some kids are foolish enough to climb it though it hasn't been maintained for twenty years and the rungs of its ladders are rusty. The hillside between the chimney and Peñarroya is full of fissures and remains of underground buildings.

On the side towards the road that leads to Cala Cortina there is a sheer drop fenced with wire. Alongside the road you see locked-off passages and gas containers. This is an area no-one can explore. It is military property.

There are many mines recorded as having operated in Santa Lucia, some of these on the slopes of San Julián. Mostly the exact positions are not known now apart from the few on the Carlos Lanzarote map. Some of the names are a little odd. Given that it is a fishing area there were mines named Tiburón and Marrajo (shark and mako shark). The latter name gave its name to one of the processionary groups involved in Semana Santa because the fisherman paid for costumes and images by selling mako sharks. The Marrajos are one of the largest groups as are the Californios who were particularly lavish with the gilding of images as their group was formed with mining money, the name coming from the designation of the Sierra Minera as Nueva California.

When I was in a small political group some of its members were in the Marrajos. They told me as I was a miner if I was interested in joining the religious processions I would have to be a Californian. Perhaps the oddest names of the mining concessions of this area are Si, Si and No, No.

In the eighteenth century, there was a quarantine hospital for the sick brought up from ships moored below. There are almosr no traces left. It also had a spring alongside which is now lost. Only a tiny bit of wall remains on one of the slopes not too far from the chimney.

A rocky road connects with that which goes up to San Julián Castle and then turns down into Escombreras past the Trincabotijas fortifications. It is a possible escape route if there are ever major problems in Escombreras. Here are the remains of the Trincabotijas batteries, mostly nineteenth century.

San Julián runs across to Monte Calvario. The tops

are connected across a ridge which provides a pleasant walk. It was a little difficult to find until someone marked it with blobs of blue paint on some rocks.

All of this territory runs above or beside the notorious district of Lo Campano. Lo Campano is a boil on Cartagena's bum. Rumours on the grapevine say that all the illegal houses there (a great many of them) will be razed to the ground and a golf course built. Spain with its drought problems does not actually need another golf course, but anything is thought to be better than Lo Campano. I used to visit it regularly to buy empanadillas and a tart filled with cabello de ángel from an excellent baker's. Cabello de ángel, angel's hair, is a sort of jam made from pumpkin or sometimes custard apple. While the baker and his family seemed to be first class people, those on the streets looked a little strange, rather like extras from a remake of the Night of the Living Dead. I eventually stopped my frequent visits there when I was questioned by the police and my handbag searched. The officers concerned seem to think I might have been buying something other than empanadillas, even though I had a carrier bag full of them as my alibi. The baker's has now closed down. I hope this hard-working man and his family found work in a safer part of the city. A building with a workshop has now opened a little further up the road. They are struggling to improve conditions locally by providing courses for the unemployed.

The La Verdad newspaper carries frequent stories of goings on in Lo Campano. Garitos (drugs dens) are raided and razed to the ground. But, like the Hydra's heads many more spring up in their place. There was a picture of a home-made assault rifle. No damage seemed to have been done with that. Perhaps its owner was too stoned to aim straight. One successful murder there was of a Moroccan who was hit over the head with a baseball bat for peeing in the road. I have asked my husband never to go jogging in Lo Campano in case he gets taken short...

If I talk about Lo Campano people invariably say, rather nervously that its inhabitants are "buena gente", good people. It reminds me of the way the Ancient Greeks called the Furies the Eumenides to appease them. There is some truth in it though. Some are struggling to improve the area with book exchanges and courses. One Lo Campano lady I met was a regular demonstrator to save antiquities. She reminded me of a Greenham Common woman, singing *No nos moveremos*, the Spanish equivalent of *We shall not be moved*, with glee.

The best bit of architecture in Lo Campano was the long disused Art Deco Naval Prison in the pinewoods. I'd like to have taken a closer look and photographed it but it was razed to the ground without protest or warning before I got around to this. It had been yet another meeting point for dealers. There was a huge drugs bust there. The local paper said that all the prison doors had been stolen as the owners of the garitos wanted reinforced metal to keep the police out. Kids used to play there and the danger to them was given as the reason for demolition. Local rumour speculates that it was another sacrifice on the way to make a mega development, a clearing of a huge piece of land. It was an interesting building that could have been made into a posh hotel to go with the golf course. At one time, reopening it as a general prison was mooted but there were local objections, so the idea never progressed. This was in a period when Nueva Santa Lucia was being built with new blocks of flats in an attempt to upgrade the area.

Lo Campano has cheap good quality drugs of every kind. I know this because my son's classmates have told him so. Rumour has it you can buy absolutely anything in Lo Campano... magic mushrooms, cannabis, cocaine, heroin, explosives from a local quarry... No problem. A small amount of cannabis is legal in Spain, but growers here usually exceed the limit of one plant per adult and have a whole plantation. Heating is often obtained by tapping into the street lights which keeps the electricity bills down. Occasionally police helicopters fly overhead and find these hotspots with special heat sensitive photographic equipment.

Lo Campano is set in beautiful countryside. I would love to take walks there. I have on a few occasions when I gritted my teeth and made my way past the zombies on the streets. I was rewarded by a walk amongst the largest rosemary plants I have ever seen in my life. When there's no money left for drugs, the locals smoke rosemary...

Lo Campano is at the end of the Sierra Minera where the mining country runs out. There is still a chimney set on a small hill with mining remains around. The remains at the back of the chimney are currently being squatted. A road and a path wind up Monte Calvario from a turning close to the cemetery. There are lines of foundations in wasteland above the graveyard that may have belonged to an earlier monastery. The graveyard itself has one or two impressive tombs. Victor Beltrí designed some of them and a group interested in his work puts on occasional guided tours. I have been on one. They raise small donations for local charities. The cemetery also figures in the rare legend of the vampire of Cartagena. This story has been made into a novel by Fernando Gómez. It tells of the arrival of a coffin in the port of Cartagena. The coffin is then sent on to La Coruna and various nasty things happen in the towns it passes through. Eventually the coffin was returned to the city when no one claimed it and it was given burial in this particular churchyard in an unmarked grave. It's an enjoyable novel but is very short. I wanted a few more stories once the characters had been established. On the whole I feel slightly safer in the graveyard than other parts of Lo Campano, but then I have not been there at night.

A road runs between Lo Campano and Escombreras from halfway up that to Calvario. Several centuries ago, the Franciscans had a monastery in Lo Campano and there are still yearly processions to the sanctuary, which is a comparatively modern building. It has a good spring alongside. We often fill up a few flagons with this holy water as it tastes better than the chlorinated version which comes from our taps. The area is beautifully kept. There is never a scrap of rubbish and the gardens are well-watered. It's a peaceful spot that gives views in all directions including parts of Escombreras that aren't seen easily from elsewhere. A small colony of cats lives there and appear to be well-fed, Maybe the caretaker looks after them. It's a lovely peaceful spot to sit and enjoy the views. There are dozens of abandoned mining wells and small buildings on the edge of small quarries that are still to some extent in use. In the manner of many dealings in the area, quarrying limits are often exceeded. Much is being lost by way of old paths and remains.

I had thought of this area being a safe and tranquil part of Lo Campano until recently when the body of a young man with several bullet wounds was found not too far from where the road to Calvario branches off to Escombreras. He appears to have been the victim of a settling of scores in the drugs trade. No-one has been caught for his murder. His family or friends occasionally leave flowers and tins of soft drinks at the place where he was killed.

There are many paths that start in Lo Campano.

I sometimes take one that passes behind the new Santa Lucia hospital. The area is known as Barranco Feo (ugly ravine) but Feo, curiously, is a corruption of Orfeo (Orpheus). This area used to be rather beautiful, with crops of pomegranate trees and olives gone wild and small hills and valleys and pinewoods and a stream of sorts. Sadly, it has been plagued by a series of forest fires set by an arsonist on windy days. Perhaps he hopes it will blow as far as Escombreras and cause a real tragedy of a dramatic kind. The stream is dry now since its reeds were burned down. Paths lead up to an area behind Vista Alegre and eventually to a triple crossroad with paths to the Sierra Gorda fortifications, various pylons and down into Alumbres.

16: Barite

13: Barite

Galena has all the heaviness of lead. A few chunks fill up your bag quickly and weigh you down. But there is an even heavier mineral in the area. Its name comes from the Greek word for heavy, barus. Cartagena is known in the mineral world for its many forms of barite. Barite often exists alongside manganese, which only became useful in the twentieth century. Barite is used, amongst other things, in the manufacture of paper. Every year the huge mineral fair at La Unión has many beautiful specimens of barite. They are snapped up by collectors from elsewhere. Many locals are blasé about it. There's almost too much of it to hold their interest. Most of us are only interested in it in its rare and more delicate forms or colours.

In the hills by Alumbres, and in the San Camilo mine in Vista Alegre, you can find white crystalline barite. This mine was beside a water tower, now demolished, and has small eroded steps down into it. Inside, it is extremely hot and airless, more so than the other mines I have visited. I have not yet explored it all. Some good specimens were found here decades ago. Some had fluorite alongside barite. Mina San Camilo is just a short walk from the N332 or the station there. In the Calblanque Park, about two kilometres along the road that leads to Portmán, lie the remains of Mina Teresita. There is a cottage-like building above the mine. Here can be found blue crystalline specimens. These need extensive cleaning with acid but can look good with some work.

I have been in the mine, but it is narrow and difficult. My son dropped his headlight down it when he was about eight and I made the mistake of jumping down a few feet to retrieve it. We had rope tied to a tree but once I tried to pull myself up by it I found that my arms didn't have quite enough strength to support the weight of the rest of me easily. I spent a worrying twenty minutes or so puzzling what to try next. Eventually I managed to get out by wedging my elbow in a hole in the side wall and pulling up at the same time. I was so glad to regain the air that I didn't notice till I got home that all the skin had vanished off my elbow. Henceforth I shall only look for blue barite in the ample spoils heap outside the mine. The samples there are rather better than what is inside, anyway. Curiously, this is often the case with mines. But the macho desire to excavate your own pieces often makes you ignore the beauty lying at your feet.

The other blue barite comes from Mina Jorge in Portmán. It is also known as La Carpeana This is the mine where Virginio Moreno was killed, No-one from the mineral fraternity ventures there now. It's not so much fear of the danger. It is left alone as a mark of respect. Across the Calblanque Park in the area known as Campos de Golf after a series of developments built by Portmán Golf, many mines also have barite, but the quality is not all that good. Some mines are buried under the houses now. One or two have their own mine in the basement, which must be fun if the owners like minerals. Failing that, a mine is probably useful as a wine cellar. Slightly more fragile barite occurs on the hills between Portmán and El Gorguel. Most of it is black and white. That of Mina Obdulia on Monte Laberinto is crystalline. Barite also occurs in the Santa Bárbara mine in the hillside above the tiny beach village of El Gorguel. The barite there is particularly delicate like white snowflakes jutting out of a rocky matrix. Some of the best barite I have found has been high on a slag heap not too far from the San Timoteo mine in the Rambla de las Colmenas. It is black and white and rather better formed than much you see in other places. Most of the crystals are unbroken and it is also fairly clean.

In El Estrecho, close to Llano del Beal, there is a mass of black and white barite with varied forms that remind me of sponges and corals. Some orange barite also turns up in this area, but it is rare. In the Haití mine there is a green form, but we have not yet found it. It is a labyrinthine mine and we have only visited a small section of it, which was deep and difficult with tiny steps which had to be negotiated by holding a chain stapled to the rock. There are some delicate specimens there set in limonite. Most are high up and need to be got from a ladder, which adds to the difficulties. They are fragile too, easy to destroy on your way to the surface. We visited it with a friend from the mineral fraternity who was strong enough to carry a three-piece metal ladder on his back on the way down.

I embarrassed my son by nervous heavy-breathing all the way down. It was the hardest of the local mines I had visited. In time I got used to it though and my breathing became normal again. Chains stapled to rock are actually a very safe way to help yourself down narrow steps. They just look seriously unsafe. Maybe there should be a few more of these around in mines and on clifftop paths.

17: Barite, Cartagena

14: Monte Miral

In the part of Monte Miral that is accessible from the road running from La Unión to Los Nietos, we encountered one of the easiest local mines. Mina Precaución. It was mined chiefly for iron and manganese. It is known amongst mineralogists for hemimorphite, a stone that seems comparatively rare in the mines in Spain. Because of this we have always found it easy to swap for other specimens when we visit mineral fairs. Hemimorphite and smithsonite were lumped together under the label calamine in mining days. It is set in a matrix of limonite.

Mina Precaución is also known as Mina San Aniceto or Mina Niño Jesus. It received the latter name because of a small image of the baby Jesus that was once there. The mine is visible up a short path from the side of the road in El Estrecho de San Ginés, just past Llano del Beal. This mine was chiefly exploited for iron in the Nineteenth and Twentieth century. It was owned by the Sociedad Anónima Minera Los Pobres and for a while was leased to Hilarión Roux, the owner of many parts of Escombreras.

For its last few years, in the middle of the Twentieth Century, it was owned by Minera Celdrán, S.A. Cost of extraction and transport by then was higher than the value of what was being mined and it was abandoned.

When I went there, half a century later, there had been a half-hearted attempt to fill in the entrance with breeze blocks. Someone had completed a few rows then abandoned it. Were they trying to make a house? Or was it intended to stop people going inside the mine? As I looked down into the cave, I could see that it was in a good state of preservation. I came back later with helmets and lights and my son. It was the weekend and the weather was cracking up. We were glad of the cave to shelter from the rain. It was the first mine we ever visited. We were not alone. A party of men emerged carrying boxes of rocks. They were local rock-hunters they explained. Possibly they were members of the society we were to join later. We have been back there many times since.

The cave was warm and had a lower part that was reached by a flight of broad steps. All was intact on this lower level and looked as snug as the cellar of a house. Some calcite glittered in the walls. The best minerals were in the upper part though, glittering hemimorphite. Many good specimens were lying on the ground. It was an exceptionally easy mine to start our explorations. There are micro minerals alongside the hemimorphite that are also worth collecting. I did not learn this until much later. You would need a loupe to see them. Amongst these are chalcophanite, hydroheterite and hydrozincite.

The rock of this mine is very dry and porous which means it can be strongly affected by water. Years later this lower area became much more dangerous. The ceiling fell after a period of heavy rains. At the time, the upper area seemed precarious too and the noise of the local trains echoed through it more loudly than before. Surprisingly, it stabilised when the rock dried out again. Hemimorphite in a matrix of limonite is one rock I never wash. The stones from this mine soak up water like a sponge. I find a little hope in nature's restoration of the mine. All is not lost. Not everything is deteriorating and disappearing. In some other cases too, the removal of minerals or rockfalls expose other treasures. The minerals we seek are not about to run out although the spots we go to may change from time to time. I learned recently that the parish of El Estrecho used to perform mass in this mine at Christmas. I would rather have liked to have seen that. I suppose that Mina Precaución is one of the few mines where this would have been possible with its easy access. It's far less dangerous and cramped than most mines in the area. You couldn't ask an elderly congregation to crawl along a cramped passage before enjoying the Mass.

Amongst the hills riddled with mines that make up the Sierra Minera, Monte Miral is perhaps the strangest. At the foot of it lies Cueva Victoria. This is a huge cave complex two kilometres long that runs beneath the mountainside. It is usually closed off, as it has been the subject of a long ongoing archaeological excavation. Early samples of human bones and animal ones have been found there, an exotic array of the latter including mammoth, rhino, hippo, sabretooth and hyena bone fragments.

There seems to be evidence that these and deer sheltered in the cave If you walk down the steps to its grilled gate you can glimpse its dusty interior. One year we managed to take a trip over it, the only public opening of the cave in three years. There are occasional tours, which always get booked out quickly. Only snatching a chance got us there. The clocks had changed on the day, and nobody was at the station on time, the scheduled meeting place. We hastily made our way to the cave itself, armed with helmets, and were assumed to be on the list. We were taken over it by a group of cave rescue men with carborundum lamps on their heads. These are very efficient and give a much stronger light than electric torches, but the smell made me feel ill afterwards. They were in common use in the mining industry as the mines of the area do not have a methane problem which could ignite. The complex of caves is so immense and so labyrinthine it would be easy to get lost without a guide.

Our trip took two hours, but we saw only a small part of the caves, varying from domed chambers to long passages. In the Roman days, we were told, there may have been thermal baths here. But it was occupied long before the Romans, as the bone fragments attest. It is claimed that a human bone found here dates back to eight hundred thousand years ago, or a million years according to others. This bone belonged to Homo habilis who was not the brightest sample of prehistoric man. His weapons were exceptionally primitive and only up to the most basic hunting. The sort of animals whose remains ended up in there would definitely have been out of his league. The guides talked of the dangers of drinking cave water and now I have become more careful than I used to be. They also talked of cave viruses. Mines hadn't made me ill before, but I was hit with flu-like symptoms after this trip. I suspect the smell of the lamps affected me more than the cave. Recently, I went to a talk on Homo habilis by Professor Javier García del Toro at the closed gates of Cueva Victoria. The path had deteriorated further, there was no rail to the steps and the inside excavation has been abandoned because of safety issues. It is yet another of the excavations which desperately needs serious funding.

A legend has the hill above depopulated in the eighth century after a huge forest fire. The flames were so hot they melted some metals, facilitating the discovery of various minerals and processes. This is similar to older legends from other parts of the world like that of the "Copper of Corinth". Various amalgams of metal have probably been discovered this way, though I suspect most of these discoveries were far older than the eighth century.

Above ground, the mountain is no less strange. If you walk round either side of Monte Miral you will see several small buildings on its slopes. Most of these are not the usual mining buildings but are in fact hermitages. Across the La Manga motorway at the foot of the mountain lies the beautiful, but ruined, Monasterio de San Ginés. Its buildings are from the sixteenth and seventeenth centuries. It is now owned by Hansa Urbana, a building firm. The original plan was for Hansa Urbana to restore it as part of a development project. It was for many years a finca after the Franciscans departed in the nineteenth century. Little was being done to preserve or restore its former glories, but a small group has formed to fight for its restoration. It was built on the site of a thirteenth century monastery. The area had been a place of pilgrimage for many centuries before. San Ginés de la Jara is usually identified as St. Jean of Arles. It's a long-standing cult which involves in one version, a miraculous transition of the bones by other saints to this spot...

Nobody knows the exact location of these relics. The spot is mentioned as holy by both Arab and Christian writers. The eleventh century Chronicle of Al Himyari refers to it as does the twelfth century Codex Calixtinos, which mentions the transfer of relics from France to Cartagena. Probably some sort of holy building from the Visigoth period was converted to a Mosque under Arab occupation, returning to Christian use later.

At one time there was a funerary stone in the walls marked with C. Numisius. This was stolen in the period of Hansa Urbana's ownership and other parts of the building have suffered damage and expoliation. It turned up again buried nearby and for a while was in the little museum of San Pedro before its provenance was understood. It is now going back to the Archaeological Museum in Cartagena at some stage.

A recent attempt to mend the roof used modern tiles and the restoration work lacks authenticity. It is a sad state of affairs. The Roman stone may have come from a previous villa on the spot. Perhaps the Numisius clan were the developers in ancient days. The same name occurs in other stones in Cartagena's Archaeological Museum. The monastery also suffered a great deal from developments in 1931, when it was turned into a house and floor levels were altered. There is also evidence in its stonework of a link to the seventeenth century coastal defence structures made against the Barbary pirates in the reign of Philip II. San Ginés is still celebrated with a romería in August, but the separation of the hermitages and the motorway close by means that old routes can't be used. An entirely different route is used for a small procession with donkey carts and walkers who proceed from the Iglesia de Caridad in Cartagena via La Aparecida and ending up in Roche. Sausages are eaten in La Aparecida and paella at the end of the pilgrimage. The nine hermitages on the hill were dedicated to saints who could never have lived there. A German stone collector I met at a boot fair told me that the Monastery was also the site of unofficial guided tours by psychics who concentrated on the crypt area. He told me that twenty female skeletons had been found there, a little strange for a monastery... "I don't know what they did to those poor women," he said.

One curious legend places the tomb of a female martyr on the site of the monastery in earlier days. Her tomb was topped with a cupola. No bird could fly over this without being pulled by some unknown force into it. In 1023 or 1024, a group of French Christians came by boat and collected the remains from this mausoleum.

In El Algar, nearby, there are local legends and ghost stories concerning a series of galleries or underground passages connecting the monastery with the town. Algar means the cave in Arabic. The purpose of these seems to have been defence against the incursions of Berber pirates who were a constant threat from the sixteenth to eighteenth centuries. No knowledge of the exact position of these passages seems to be left. Perhaps old mine workings were a part of them. YouTube is full of short films where psychics visited the Monastery and claims are made on these of "psicofonias", voices picked up on recording machines at key spots. One seems to be snarling: "cerdo" (pig). Claims have also been made that the monastery was among several out-of-the-way spots used for secret tortures by the Inquisition. Another was the finca known as La Buena Muerte in la Aparecida.

Fray Melchor writes of the area being mined by Phoenicians, Carthaginians and Romans. Most others see the Phoenicians as traders rather than miners. More likely the native Iberians or Tartessans mined there and traded with the passing Phoenicians. The name of Monte Miral has been suggested to be a corruption of either Mineral or Milagro (miracle). Other possibilities are mountain of the river as a corruption of Arabic, or Mirall meaning mirror in Valenciano. There is no river now but there might have been in past ages. More water there perhaps, if the speculation about a thermal establishment in Cueva Victoria in Roman days is true. In one instance Mirab instead of Miral is used. In view of the Arab connections this could have yet another meaning. Mihrab is the name for a gate shaped niche pointing towards Mecca. I am always conscious when visiting the hermitages of their outlook and views through windows etcetera to other places. In the fifteenth century and in the writing of Fray Melchor de Huelámo (1607) other names have been ascribed to the mountain: Laurinum, Laures, Larim. The last word sounds like Arabic but I couldn't find a meaning. There is a people and language in the Sudan called the Larim. In Persian it is a place name and a silver coin. Could the coin name have any connection with Laurion in Greece, which was famous in ancient times for its silver mines? I have passed these outside Athens. They still exist but are no longer mined. The local magnate, Hilarión Roux, had a hand in those too, though I had never heard of him at the time I visited them. The name Laurion appears to have come from the Greek Laura for a straight narrow road. The mines have a straight narrow tunnel, from which, these days, foul air issues. Interestingly Larim is Miral backwards. Which could mean the name arose out of a misreading of an Arabic word. Could Laurinum or Laures be a corruption of the Greek, Laura? Or could they have any connection with the Roman cult of the Lares? The Lares were the family ancestors worshipped by many Romans. Family is important to Spaniards too, more so than to many modern nations. Me cago en tus muertos, I shit on your ancestors, is the strongest level of Spanish oath. The Greek Laura has another extremely interesting meaning. Laura or Lavra was, from the fourth century applied to a kind of monastery made up of hermitages. The Lavrite way of living became a system whereby hermits lived in their own cells near each other, meeting up say once a week for a communión in a central building. It seems to me very possible that the hermits of Monte Miral were just such a community, and met up in what is now the Eremita de Los Ángeles, the biggest of the buildings, or in the Monasterio itself. Probably the original hermits first occupied the area in the days of the Visigoths and their hermitages got rebuilt across the years, reusing some of the old stone. They may even have been on top of earlier temples of a more pagan kind. The Visigoths of Cartagena were mostly converted by the Saints who lived there in those days, Isidoro, Fulgencio, Florentina, Leandro, Teodosia who were of Roman descent. Most Visigoths were Christian rather than pagan, though some, like Teodosia's husband, subscribed to the Arian heresy. The four saints are still important in Cartagena's worship. Their sister, Teodosia, the fifth saint, gets left out, either because of her Arian connections or her marriage.

Monte Roldán, a few kilometres from Cartagena in the opposite direction to the Sierra Minera is supposedly named after Charlemagne's knight, Roland: Roland, Roldán and Orlando all being versions of the same name. Legend has it that he and his brother, Oliver, came to Spain in search of their brother, Ginés El Franco. Ginés was also the nephew of Charlemagne whose court he abandoned to seek out the tomb of the apostle (James?) He was shipwrecked in Cabo de Palos, a site of many shipwrecks before and after, near the monastery of San Laures (yet another version connecting similar names, similar words). He retired to live on Monte Miral for 25 years. Roland had climbed Roldán three times as it was the highest spot of the area and from there saw smoke above Monte Miral. The brothers took their ship to Cabo de Palos then in search of Ginés. Next time I go up Roldán I shall look in that direction to see if this story has any elements of truth.

The area's modern names are Cerro or Cabeza (hill or head) de San Ginés. There were supposed to be a total of nine hermitages on the hill. I have two little books on these. According to one, only six of their hermitages are still identifiable and these are falling fast into decay. Just three of these have been declared national monuments and this has left them in a kind of limbo where they cannot be restored and where casual vandals still pick away at them. The most easily identified buildings run in a line down the back of the hill. Close to the top, visible from many kilometres away, the first of these is a small square building with a domed roof. It is built in such a traditional way that it is hard to decide what era it springs from. The same is true of most of the other buildings. They commemorate an array of saints, some of them desert fathers, an inspiration for later anchorites. The second little book I have on the subject identifies each and every one of these: Saint Onofrio, Saint Paul the Hermit, Saint Antonio the Abbot, Saint Hilarión, Saint Jerome, St. John the Baptist, Mary Magdalene, Saint Francis. This book sometimes makes the mistake of identifying mining remains as hermitages as the author makes desperate leaps of faith, wanting everything to exist as it once was.

The dome of the highest hermitage is composed of small overlapped stones set in thick mortar. The building is just under 12 square metres. When I first visited this spot, the small building was almost intact and hundreds of unopened packets of crisps lay on the floor, a gift to some modern-day hermit, perhaps. When I saw it again a year later there were no crisps, and someone had been digging away at the mud floor. This was probably the work of vandals rather than archaeologists as there were no markers or tapes of the sort I would associate with the latter. The digging was a little mysterious as when I first saw the hermitage the floor was flat but trodden earth and rock rather than anything worth pinching. This hermitage was dedicated to San Onofrio, according to the most religious of my books on the subject. In time further vandalism ensued and its door was outlined in red paint and Libertad written on the wall. The small domed building is a landmark in the way the other hermitages are not. It can be seen topping the hill from many kilometres away. A hole in one of the walls has a sightline to a ruined building a little higher up. While this building is not a hermitage it is possibly on the site of a lost one. Maybe some old stonework was incorporated but nothing is entirely evident. It is more like a small mining building or cottage of the nineteenth century than anything holy.

Straight below this hermitage there is another. The building was circular and domed but most of the roof and half of the walls have now fallen. It is now little more than a cross-section of a dome. You can clamber across chunks of masonry below. It was sacred to San Francisco. Judging by old diagrams it had seven or eight niches. Presumably paintings or images adorned these apart from one that was a window space. Further down, there's the most interesting of the three buildings. It is more like a ruined church, rather too posh for the possession of one poor hermit. This was the Sanctuario de Los Ángeles. On the side, visible from the motorway, you can see the letters AVEM, the start of the phrase Ave María picked out in dark stones in the building's walls. There are many thorny plants there now and access is hard unless you find the correct path, prickly pear, agave, wild asparagus, etcetera, block your way.

Probably this sanctuary once had a marble floor as there are broken fragments outside. The paving stones are long gone now though. Again, there are signs of amateur digging. The building seems to have had two storeys as there are holes. which must have once contained horizontal posts supporting a ceiling. There is a tiny dome. Inside it has the remains of a fresco. There are several other parts of the room that links with this dome that contain fragments of paintings. One wall has the traces of a few figures, perhaps former hermits or saints connected to San Ginés, or perhaps some of the saints commemorated in these hermitages. Beneath the building there is a cellar, which was probably a water storage area. It is easy enough to drop down into this through an aperture. The floor is full of broken pottery.

In the fold of the mountain beside there is another ruined oratory. It looks square from the front but the walls slope behind. It is in fact a rarity in architecture, a seven-sided building. It is hard to perceive this from inside but it is apparent in the aerial view on Google Earth. It has sharply defined windows and doors. In front there is a damaged wall with a caper plant. Again, the floor is broken inside. Someone has dug away into solid black rock. Roots are poking their way through what is left of another domed ceiling. Most of the inside plaster has gone but this was probably painted at one time. There is a rim of red on the inside of one window. One of my booklets speaks of the base of an Ángel but this is long gone. This building was dedicated to the penitent Mary Magdalene. I find this hermitage a little creepy partly because of the black stone and partly because of the silver flattened heap of slag outside in the midst of which is a mining well without any kind of edge of stones to keep it safe. Across from the flattened area there is a well-beaten track that leads down to level ground.

If you wish to walk between the surviving hermitages it is easiest to start with this one as a hunter's path leads up to it. If you exit the building on the left the path continues up to the Sanctuario de Los Ángeles and the two hermitages in a straight line above. From the top, another straight line takes you down to the only other remains that can be found at the present day. The ruinous cell of San Antonio contains large chunks of broken stalactite from a mine nearby. The other hermitages and their exact locations seem to be lost. Probably a part of a sixth was visible not too far from this one as it is mentioned in the more sensible of my two books on the subject. Those remains have now gone but were probably on this line further down the mountain. They may have been covered or reused in other building work.

Few books hazard a date for these hermitages. One newspaper article said seventeenth century. I managed to get an answer from an archaeologist. He said that they were all eighteenth century, built at a time when death cults were popular. I would hazard a guess that some of them used older stone from earlier hermitages in their buildings. Sadly, I saw some evidence that bricks from.one of the lost hermitages had been used in one of the mining wells. There are several mining buildings on the hill that people sometimes mistake for hermitages, but they have none of the same characteristics. Probably nineteenth century miners demolished some of the sacred places when they got in the way and old stone was reused. The miners of those days were mostly religious, and it seems probable that they were forced to desecrate these buildings by their overseers.

There may in fact be a much older origin to these holy spots. Professor Javier García del Toro has mentioned the existence of some Roman stonework at the base of one of the hermitages that was exposed in old excavations of the 70s. Perhaps there was once a series of small temples whose use was changed at the coming of Christianity. As a person with pagan leanings I would like to believe in this older origin. More excavation needs to be done. A preservation order has been put on some of the hermitages now which all lie on land owned by Portmán Golf. But so far only three are on this order.

As well as the hermitages I have seen a photo of a cave near the Sanctuary. This is now impossible to find. I have walked amongst these hermitages many times alone and once with a group of people who are trying to save the monastery. The local artist, Javier Lorente, is amongst their number. He was brought up in one of the houses attached to the monastery when it was a finca. He has a wealth of anecdotes about the area and it figures in many of his paintings and photographs. When he was a child there was a story of buried treasure guarded by a serpent. Nobody discovered the treasure, but a large snakeskin was found on the property.

I have urged people from this group to get the hermitages included in any plan for restoration. I am friends with some of them on Facebook now and one posted a picture of the top hermitage on Hallowe'en with a solar alignment at one in the afternoon. Was this significant? I have sometimes speculated that there is some kind of numeric mystery connected to these buildings, but I have not worked out what it is... nine hermitages, a seven-sided building, another with seven or perhapd eight niches, a fresco of six saints, etcetera. Are there other alignments with the other buildings? Perhaps one day I will find the truth.

My son had not visited the hermitages and I took

him one day. At this stage I knew only the three that are in a line with each other. After visiting the upper ones, we sat down and rested at the larger hermitage de Los Ángeles. We had eaten our sandwiches rather early and it was lunchtime. Was there any chance we could get into Chupa's? Possibly not, as we had not booked. I said I would ring when we got down the hill and see if it was not too late. Son was determined to leave by a different route and I shouted at him to wait at the bottom of that area. The wind must have carried away my words. I waited and waited at the bottom, conscious that the route he had insisted on taking was rather hazardous.

Could he have fallen down the mining well?

When he did not reappear, I walked up to the main road in case he was waiting there. After no sign, I decided to ring the police. It was not too long after the famous disappearance of Madeleine in Portugal so they took it very seriously. Eventually, he was found and brought back to me at the stage when they were thinking of sending a sniffer dog up the hill. He had devoured a leisurely paella at Chupa's and viewed the mineral collection of his neighbour, "Sevillano" a friend. The neighbour had been in the local police, and so heard the missing kid bulletin being put out locally. He drove us back to Cartagena and so all ended well. There were a few half-hidden smirks amongst the rescuers. I think they liked the enterprise of a 9-year-old English boy blagging his way into a free paella in a Spanish village while his mother fretted a couple of kilometres away.

James was in Andalusia at a chess tournament and I kept the story from him for a while. My son's adventure ended up in La Verdad next day and was also soon known in El Descargador as one of the staff was from Llano del Beal. It's a small world. The next day I visited Chupa offering to pay for the meal my son had. He wouldn't take my money and insisted I have a glass of wine and some cheese and biscuits as I looked tired and cold. My son insisted he had not been lost but that it was I who was lost. I had had no lunch and hung around worried while he was downing paella in comfort. Somehow what happened reminded me of the ambience of displacement in that area of the mountain and its dedications to alien desert saints, the moving of relics, the fading outlines in the frescoes, legends of missing treasures, the unknown... Catholicism has many magic elements with its stories of flying houses of the Virgin, multiple relics and miracles. These stories have less appeal to the modern generation reared on science, but the stories still exist in old histories.

At the very least their symbolism has some mythical value. Some places attract such stories, while others do not.

18: Fluorite

15: Fluorite

In the sixteenth century, alum (alumbre) was discovered to the east of Cartagena. The settlement built around this lucrative trade was known at first as Alumbres Nuevos (New Alumbres) to distinguish it from Alumbres or Alumbres Viejos near Mazarrón. The chronicles of the time show the existence of an ancient building. Years later it became the inn, belonging to the Duke of Escalona and Fajardo family, the Marquis of Vélez and Molina. Their major domo, a man called Mínguez Inglés, better known as El Rico (the Rich Man) lived there with his beautiful daughter. There were frequent attacks in the area by pirates. They kidnapped her and demanded a huge ransom from him of the alum he had not yet delivered to his masters...I had never before reading this thought of alum as a commodity worth stealing. It has been used as deodorant, in medicine and to fix colours in dyeing. It is undoubtedly useful but seems like a cheap article in itself.

There were already mines in Mazarrón and Rodalquilar in Almeria and a demand for alum in Flanders and England. In 1534, a royal decree allowed the cutting down of the forests and use of the waters necessary for the working of alum in the bishopric of

Cartagena. In the early days there were only about a hundred miners, single men, working alum and some lead. They put in a request for some land to till. Their security had to be guaranteed by the mine-owners with the help of Cartagena's council because of the incursions of Berber pirates. This population was reduced by epidemics to only 20 residents in 1561, although two years later the church count rises to 205 people and 45 houses. In this same year there was a large invasion of Ottoman Turks into two bays at Algameca, the other side of Cartagena. This was subdued with the help of troops from inland. From the year 1575, the increase in the Barbary corsairs significantly reduced mining activity. In 1584, 300 Turks came to land in Calblanque, further along the coast, and roamed the area until June. They went to Alumbres to board galleys in El Gorguel after taking many captives. Facing a new threat on 4 February 1589 the City Council ordered that the residents of New Alumbres do their part in protecting the area. In 1590, thirty soldiers were put to guard the coast, in 1592, this was raised to three hundred. By this stage, there were again only twenty inhabitants and the mining died out and was not revived until the nineteenth century.

The men left made a meagre living by picking up lead and esparto grass. A trade in ochre had begun in the seventeenth century. While I have never bothered to collect alum in Alumbres I can see the presence of ochre by the soft reds and yellows in the earth alongside the quarries between there and La Unión.

There were three flour mills in the eighteenth century. Two still exist. One is beautifully restored. I think it is a private house now, rather than a working mill. Nearby, there's the ruins of another. In the nineteenth century Alumbres was incorporated with the city of Cartagena. The town grew and soon had its own cemetery, theatre and soap and explosives factories. The cemetery contains the remains of many cholera victims. The explosives factory was owned by Miguel Zapata, otherwise known as Tío Lobo, of which more in another chapter.

Alumbres was supplying Cartagena with water at the beginning of the twentieth century from its wells and a large water deposit in the La Parreta area. Today the La Parreta area is only known for its extensive ruins of industrial architecture. In the old days Alumbres had many different barrios and included an area from La Esperanza to El Gorguel and Vista Alegre. By 1920 it had 1396 buildings and more than 4000 inhabitants.

If you walk around Alumbres you can see that it was once a charming village from the intact windmill on a hill on its outskirts and the pretty countryside full of wild flowers. In the summer there is a fiesta of San Roque, its patron saint. The first time I heard of San Roque was in a famous Spanish tongue twister:

El perro de san Roque no tiene rabo porque Ramón Ramírez se lo ha robado.

San Roque's dog has no tail because Ramón Ramirez stole it.

San Roque otherwise known as Saint Roche or Saint Roch may have started as a saint who was prayed to against the tempest, or for cures to the plague as he had treated victims. There were theories in medicine of the middle ages that wind caused plagues so the two things could be connected. I loathe wind and have always felt less well when it blows strongly. I mentioned this to a friend with Chinese blood and she explained that wind lowers your chi thus making you vulnerable to infections. At one stage, San Roque was brought food by a dog, thus the dog legend. In some versions of the story San Roque had bubonic at the time the dog helped him, thus saving him from death. San Roque is also seen as a protector of dogs therefore. There is even a dog shampoo named Saint Roche. Nearby Roche, which was once a part of Alumbres, must also be named after him.

Cartagena itself had several epidemics of the plague: 1348, 1558, 1648 and 1676. Ports are vulnerable places when it comes to infections. The plague came down the coast from Valencia in 1647. There was an edict at the time banning anyone from treating or dealing with any victims who turned up in the neighbouring towns from Alumbres to Fuente Alamo. There was also a veto against the trading of clothes and furniture from without the city. Bread ran low due to a ban on the wheat trade with neighbouring towns. A boat from Genoa and another from Naples were quarantined at Escombreras with orders to the locals not to allow the crews to come ashore. The image of San Roque was processed from its hermitage to the church of San Domingo on the 2nd of November. San Roque was prayed to against the plague and also for water. At this stage, maybe thinking prayer was enough, some inhabitants on the outskirts of Cartagena took inhabitants from Valencia and other infected areas into their houses. The mayor ordered these houses sealed for the time being. In spite of all precautions the plague reached Cartagena in 1648. This particular epidemic was the worst and halved the population together with the effects of hunger from the near siege-like conditions. The population in Cartagena was at that time lower than that of the nearby village of Alumbres.

Epidemics were often a problem in Alumbres also, and it is certainly not a healthy place now. It has been cursed with a large number of cancer and leukaemia cases. There are odd lines of graffiti in the town blaming Repsol, the giant oil company. Forums also talk of the number of pylons and power lines running through the area. Maybe the modern-day inhabitants of Alumbres should pray to their saint for some of the local industry to close down or become safer. They make occasional pleas for a green belt of trees to separate them from the heavy industry, but little has been done. I went on one of the tree planting days and duly stuck the small pine I was given in the ground. Repsol is conscious of its image problems and also does the odd ecological day in schools. My son was given a little set with earth, pots and pine and birch seeds. But trees are very difficult to grow from seed and we didn't manage it. Some more effective tree planting days would be a good idea. Why give someone just one tree to plant when they could do a dozen or more? Repsol could get together with CreeCT or ARBA who have organised tree planting in other parts of Cartagena to reforest areas damaged by fire.

Alumbres was also scheduled to be the spot where a new Befesa incinerating plant will be built. There were local protests. Local doctors were not happy as they already had many serious problems to deal with in their health centre. A friend campaigning for a small political party was told by the local doctors of high rates of deaths from cancer. A survey I saw years ago seemed to confirm this, although recent ones appear to have less frightening statistics. I hope this is a sign that things are improving rather than some juggling with figures.

We made a few other visits to Alumbres as an enterprising bar was hiring stand-up comics from all over Spain. We had four wonderful nights there in the bar Yo Que Se. Spanish comics tend to be lewd and a little crazy in the manner of Billy Connolly or Eddie Izzard. We laughed till we cried on some occasions. Best for me was Ángel Miralles with his wonderful monologue on shitting. He used a chair to mime all the different methods. The bar has since closed. Maybe they didn't charge enough for their drinks. I had several nights there with my son on a total spend of 5 Euros, two drinks apiece, crisps for the son and a bowl of free nuts thrown in. The bar is now a fruit shop. Recently we found out that it was owned by one of the brothers who own Bodega Velasco nearby which does a high-quality menu del día.

A couple of other moments from those comedy nights stay in my mind. One of the comics was telling anti-Basque jokes. Basque jokes from Spanish comics are pretty similar to Irish jokes told by English ones. The Basques in these tend to be slightly thick or odd. Unfortunately for the comic there was a Basque in the audience who objected. He was an extremely drunk Basque. The comic thought on his feet and gave him the mike for a right of reply. Alas, the guy was so drunk he could only stick it in his ear. The comic promptly said that he must be the only man in Spain who could give a blow job with his ear. On another occasion the English were the butt of jokes. I guess the comic concerned had not expected there could be any English in the audience in a small bar in a wholly Spanish village. My son soon told him. He was deeply embarrassed and tried to recover by saying it was only English blondes who were stupid, not the rest. I was brunette at the time.

Eliécer from our mineral group had told us that Alumbres was the best spot locally for fluorite. He described a hill with wells along the road to Escombreras, drawing a map on a paper napkin in the café we were in after the Annual General Meeting. I managed to blow my nose on the map later that day and throw it away by accident, but we still found the hill the following Sunday. The road from Alumbres to

Escombreras is a nightmare on any day but a rest day. It goes down to the port through a collection of oil refineries, power stations and chemical plants. The air quality is poor there and I always take an antihistamine before venturing into that tract of country. Fortunately, the fact that all these chemical plants and refineries are in a vale keeps the pollution to one area. The range of hills either side keeps the air sweet in Cartagena and La Unión. On a weekday you can hardly cross the road for the train of lorries and tankers heading to the port. The road finishes at Escombreras but you can loop back to Cartagena via a winding coastal road that tunnels through the hills that passes by our lovely snorkelling beach, Cala Cortina.

After Eliécer's advice we searched the hill and found shards of white crystalline barite in the soil. Some of it was almost as clear as glass. As we were about to leave, we came on a large rock and detected small purple cubes of fluorite in one side. We chipped away a little but decided it was necessary to salvage the whole stone. Heavy diggers had broken away part of the hillside. If we left the stone, we would probably never find it again. I rolled it down the hillside, booting it over lumps of earth and rocks. It acquired some extra mud by the time we had it by the road. My husband came by to pick us up and I heaved the fifty-pound stone into the car. We kept it in the garage and gradually broke away lumps across the months, washing away the mud.

Months later we ran into Antonio or "Sevillano", one of the founders of our group and the neighbour of Chupa in Llano del Beal. Antonio told us that some superb specimens of fluorite were being found in Alumbres where they were broadening the road. Again, it was a case of going there away from working hours. We went down there on a Sunday. A broad mud dual carriageway had been flattened alongside the

Escombreras road. It carved straight through the hill we had taken the rock away from. We walked along it, picking through the rocks. There were large chunks of white barite made fragile by the diggers, which had powered their way through it. Small earthy caves were exposed in the cliff faces either side of this new road. We found a place where we could see fluorite glittering in the rock and dug away. We had soon filled a bag with some promising specimens. The tiny loose purple crystals went straight into my purse. It was a rescue mission as we had been warned that a few hours of sunlight damages fluorite, turning the subtle shades into a milky stone. We could see many rocks that had already suffered this fate. When we were too tired to dig for more and our bags were full, we went home. We covered the parts where we could see fluorite with mud and rocks to protect its colour. Unlike many other stones, fluorite can be cleaned with olive oil when water can take no more mud off it. Soaking it in water is not altogether a good idea as these stones seem particularly prone to developing algae, which rarely happens with quartz. Oxalic acid is something else that can be used to clean them.

The next fiesta we came back, finding bigger and better pieces. As we grubbed away at the bottom of the cliff an old man came and talked to us. He said that there had been a cave and a mine there where the road lay now. I felt conflicting emotions, sadness that a mine had been destroyed by the road makers, but pleased at our booty, so many pieces that could be swapped at future mineral fairs. He was busy rescuing a palmito plant whose roots had been exposed.

Eventually we found a vein of dark purple fluorite of a much higher quality. There was a group of mineralogists working it. We had not met most of them before although they came from Torrevieja, close to where we had lived previously. The newly-exposed purple fluorite glistened like plums in the brown soft rocks. We felt like Jack Horner hacking it out. Some of it was in cubes with the ghost of other cubes inside. It was of a good colour, but almost all was slightly damaged by the heavy diggers that had ripped through this section of the hills fracturing the ancient rock formations below ground. How many mines with their networks of galleries and ventilation shafts have disappeared for ever? We returned on Christmas day with a pickaxe. I worked off my first course before returning home for pudding with the last of the purple fluorite in my bag. We were never to find purple fluorite in Alumbres again. The road is now wider and even harder to cross with metal fences making it difficult to stray into the countryside nearby.

Everything we dug out was flawed because of the vibrations of the heavy diggers that had been used in the area. Machines of this kind cause fractures in rocks far below the surface. This was something I noticed also in Cartagena where developers damaged ancient city walls on the sides of Monte Sacro.

I made the assumption that Mina San Roque, the most famous mine containing fluorite had been destroyed during the road-building. Some easier access to it certainly was. Years later, I discovered that much of it still remained on the hill alongside the busy road full of lorries. There are several wells amongst the undergrowth there and some adventurous collectors abseil down and visit the galleries far below. It is dangerous though as the galleries run literally under the road and visitors can feel the vibrations of passing traffic.

There are other mines outside Alumbres known for their fluorite. One is Mina Marisol in the Campos de Golf the other is much closer, Mina San Camilo in

Vista Alegre just down the road. Mina Guadalupe in this area is also supposed to have had it. In the case of San Camilo, which is a relatively easy mine to visit, the fluorite mostly turned up in the waste outside. I have not managed to find any there though.

Most of the mining remains in the village part of Alumbres have disappeared. One area of Alumbres was known as Los Partidarios. The name came from the days of share-mining and small claims.

There are considerable industrial architectural remains on the outskirts, on the volcanic outcrop known as Monte La Parreta. They are a very short walk from the N332 or the FEVE station of Alumbres. It's a group of buildings that could be visited easily and could at one time have been renovated and signposted if Cartagena began to show an interest in such things. Instead they are at the mercy of scrap dealers who recently damaged the buildings. It had an interesting variety of quartz, which had crystals within crystals, though I think none in this form has been found for a long while. One of the ruins is the old washery known as La Montañesa. Amongst the other buildings is a chimney from steam machinery, Mina San Simón and Mina Manolita with their wooden headframes. There is also a small ventilation chimney that was above a powder house and the powder house nearby. One of the most elegant buildings is the San Ignacio washery with its slanting transport systems.

The miners from La Parreta had the job of collecting human remains when the explosives factory had a horrific accident in 1926. There were 95 workers at the time, men and women, in a workshop producing nitro-glycerine. A stroke of a hammer against an ebonite tube containing the nitro-glycerine caused the accident which shook the area like an earthquake and caused a rain of stones, wood and chunks of iron.

Nine workers were killed and another ten injured. Cruz Roja and the Guardia Civil and people from all the nearby villages helped the injured. A plot was acquired by the Sociedad de Explosivos in La Unión's cemetery to bury the nine workers and commemorate them. One of the directors of the factory was the Italian, Camilo Calamari. His palace, Villa Calamari, which once had a botanical garden attached, is in a ruinous state in the San Felix area of Cartagena. This was not the only accident in the factory, but it was by far the worst one.

All these buildings lie close to Alumbres and below the Sierra Gorda, which has an interesting set of fortifications. All military batteries are built with camouflage elements to merge with the local rock and the battery on the Sierra Gorda is aimed at merging both with shepherd's cottages and the mining scenery. It includes imitations of small mining chimneys.

A path below the Sierra Gorda runs from La Parreta to Escombreras, running by the Batería de los Dolores which is probably Cartagena's least-known military fortification. Here you can see kestrels lurking to catch rabbits hiding in the long-disused dugouts, a hangover from the Civil War. There are a few mining remains on the hillside above and a tiny path leading to the Cabezo de La Porpuz, a mountain whose top can only be climbed on a system of rungs. The lower parts can be walked. There was a mine there which is said to be the most likely spot to find stibnite, which is rare in the Sierra Minera.

The other mining remains of Alumbres are mostly close to La Esperanza where the border of La Unión lies. The huge walled finca there is known as Huerto de San Pedro. When the walls had gaps, I sometimes had a coffee with my husband while sitting on steps amongst the trees. We followed that by picking up sticks and pine cones for the fire at home. It is a beautiful wooded area. Apart from pines, eucalyptus, palms and an orange grove, there are many cypresses of the rare Cartagena type. Curiously there is one single one that has planted itself in what was probably once a water deposit that is the size of a football pitch. There are still broken staircases leading down into this area where you can descend to look at the sole tree. The walls are painted Pompeian red. Behind this area are the big slagheaps belonging to the Brunita quarry. From here there are many tiny paths that lead over hillocks and through woods, passing alongside old tailing ponds and dried-up stream beds. It is a pleasant and little-known area for a walk. In the hunting season part is a caza de coto, a private shooting range. Recently the walls of the finca were mended and there is only one gap if you wish to make your way in to see the trees.

19: Sphalerite (aka zinc blende, blende, blackjack, and mock lead)

16: Zinc Blende

One of the ways to Portmán is via Ruta 33, which acquired its name from the fact that it was constructed in 1933. Mining work was beginning to decline, and it provided jobs for a hundred men for a year. Its purpose was to connect La Unión with Portmán. Today it survives and the first kilometre or so is a part of the Parque Minero. A fake mining train runs up it to the newly-opened Mina Agrupa Vicenta. Fake trains are a part of the culture of tourism in many towns. They look like trains but don't run on rails.

Eventually, the further parts of the road were cut through by quarries. In my first attempt to travel this road with my son we made an epic descent via what became known in our family as "the silver stream bed". Streambeds sometimes work as pathways, but they are far less regular than normal ways and sometimes include drops of a few feet where water once cascaded. We negotiated this streambed down into Portmán.

The deal was that if we couldn't step down anywhere we would have to ascend again. I have learned not to jump down from rocks because you could get stuck somewhere without the ability to descend or ascend. The bed of the long-dried stream was silvered by a mixture of crushed slate and pyrite.

It is clearly visible from the village below. It connects, eventually, with a more formal rock-lined channel similar to the one in Llano del Beal. These channels are dry most, if not all of the year. In practice they are probably a complete waste of time and money, as the water, when it comes, has other ideas and makes its own paths where it will, sinking into the hillside and disappearing as quickly as it came.

Our second trip was by bus, but this only allowed us about three quarters of an hour before the return service. I acquired an admirer on the bus who kept complaining his wife had gone to La Manga and gnashed his ill-fitting false teeth. We ran into him several times on this route and christened him Señor Loco. I sort of got the impression the wife had never returned from La Manga. On our first trip we found a dog's skull on the beach, probably a greyhound's, bleached clean by the acidic minerals in the sand. Señor Loco had told us that there were wild dogs in the hills here. I have never seen any although I did once see a couple at El Gorguel and another couple in the Emilia quarry.

When I first started walking the area, taking Ruta 33 to the top of the Sierra Minera was a good way to get lost.

Now it is firmly signposted and connects with a path labelled MU88, which leads down alongside the San José quarry and on to Portmán. It is about half a kilometre less than the winding switchback road used by cars. There are now two buses a day to and from Portmán. At that stage there was only one at a viable time.

I was in love with poor tortured Portmán from our first visit. Portmán, once the Portus Magnus of the Romans, suffered a terrible ecological disaster in the twentieth century. It has been labelled the most toxic port of the Mediterranean. Greenpeace closed down the mining operations by the dramatic act of volunteers strapping themselves to the waste outlet that was pouring waste into the sea from Lavadero Roberto. It was a brave act as the waste was said to contain everything from heavy metals to acids and cyanide. The local library gives away books, from time to time, and I pick up anything that looks interesting. One of my finds was a history of zinc in French. This opened my eyes to how complicated the processing of zinc was compared to the metals mined anciently. Many of those could just be smelted with wood or coal. But zinc has a low boiling point and would just evaporate with such treatment. The need for a purer form of zinc demanded much more complicated processes. The modern methods contained ever greater amounts of contaminants. My French book has diagrams of all the machines needed. Some of these still exist in a wrecked state in the old abandoned washeries. The existence of zinc as a separate metal and element was not found out in the West till the eighteenth century. It had of course been a part of brass for thousands of years before. But the ancient methods of making brass for armour and jewellery involved heating zinc minerals and copper minerals together rather than using the separate metals. The Sierra has a lot of zinc blende, which has a high proportion of zinc as well as traces of other metals, which can be separated out at various stages of heating or dissolving the ore. Lead, silver, cadmium and even minute traces of gold are in the ore. The Sierra also has the slightly less rich calamines that can be used.

    Before its modern uses as a metal, zinc was well-known in medicine. Zinc sulphate was an aphrodisiac and must not be confused with zinc sulphide, a rat poison. Some zinc is necessary in the human diet and is considered especially necessary for men. Those whose testicles do not develop properly have been shown to have little zinc in their blood. The first time I heard the word calamine I was a very small kid and had fallen in a bed of nettles. Calamine lotion was applied to my whole body to relieve the stings. A cream with zinc oxide was also popular at the time for soothing sunburn. It has an opaque quality and is still used as a total sun block in Australia.

In modern times, the need was not for calamine as much as processed purified zinc in its final form. When they hear the word, people think of galvanised zinc, which is rather unglamorous. A slightly higher grade of zinc was used in the architecture of Cartagena and can be seen also in Paris and Austria amongst other places. It combined well with Art Nouveau styles and proved a strong, practical and lasting material for roofing, easily made into overlapping tiles which could be formed into sloping roofs or stylish cupolas. Occasionally, also, it was used for reliefs and plaques to decorate the buildings. There is even one house in Cartagena, which has been entirely covered in zinc. It is not particularly pretty and has been scheduled for demolition. This is a shame, really, as it is unique in its way. When I went to a lecture on the uses of zinc in architecture, the architect giving it said this house was the only one of its kind in Europe. He was obviously itching to restore it, or at the very least, look behind the façade to see how well it had preserved the masonry behind.

The zinc has slight pleats across it to mimic the edges of building stones. The house is in Calle Caridad on the opposite side to the church and on the corner of a square the council wants to gentrify. Zinc can do most of the things lead can do in architecture and it is easier to handle, being light.

Zinc also has a history in art. Pigments containing

zinc are less toxic than those based on lead. Zinc white came to replace flake white, which was the lead-based colour. Lithopone was the name of a zinc pigment.

From the French book I also learned the word Zamak, this is an amalgam of metals including zinc, magnesium, copper and aluminium used in the car industry and elsewhere. With so many uses for zinc or its amalgams, there was a big demand in the modern world and a corresponding rise in contamination.

For decades, council agreements with the mining firms required less and less dredging of the port and this had greatly changed the face of Portmán. What was once a deep bay had filled up leaving a waste of acidic sandy sludge. The fishing industry was dying off. Fishermen were offered some compensation and jobs in the infamous Lavadero Roberto. Some took this while others moved away. Their livelihood was gone.

Greenpeace took Peñarroya to court. But Peñarroya quickly liquidated its assets and sold a large part of the Sierra Minera to Portmán Golf. It is said that the price was a peseta a square metre. They needn't have bothered, as the courts exonerated them. What they had done, terrible as its results were, was within the law. The agreements with the local town halls had simply not required them to take greater care with what they released into the sea or made them dredge as frequently as in past decades. Portmán is still polluted and the arguments still rage on. Some money has been forthcoming from the European Unión. Most has been spent in initial work like surveys.

It could be argued that Portmán Golf got the land so cheap that some financing of the work from them would be desirable. Initially I thought that the Rothschild family should be asked for a charitable donation as they financed Peñarroya and profited from it. But the Rothschilds were only involved in its finance in the early days, I found out recently, long before the serious damage was done. Probably the only hope for Portmán is from further European grants or private investment.

The University of Murcia has been part of the project, testing ways of rejuvenating the soil. A large part of the beach was fenced off for these tests. This wasn't the case when we first went there. Even without help the land is rejuvenating slowly. Part of the toxic sand is covered with reeds and sea birds nest there, or did until a recent fire burned down the reeds. Some success in salvaging damaged soil is being obtained by mixing it with marble dust and pig purine, then planting it with certain local flora known to have a rejuvenating effect like bituminaria bituminosa, a member of the clover family. After a fire on the beach destroyed most of the reeds, there were the usual rumours circulating about someone wanting to build there. It was a time of high winds, so the fire spread rapidly. Every accident or act of arson triggers these rumours about building.

Realistically, I don't think the contaminated beach, with all the publicity it has had, will be a possible spot for building for a long, long time. If it were dredged deep again there is another possibility according to archaeologists.

There have been many ancient wrecks along the coast, and some may lie buried deep beneath the toxic waste.

Portmán Golf, as its name suggests, was not a natural in the mining industry but was chiefly interested in building. It bought Peñarroya's assets during the property boom and it was much easier for them to close down all the mining operations, one by one, and make money where they could by putting in planning applications for golf resorts and urbanisations. These exist on the outskirts of Portmán.

Property prices were high, and agents held out dreams of a restored beach that has not yet come to fruition. It may happen eventually, but the authorities involved are still wrangling about which plans to accept and not enough funding is in place. Unrealistic ones of luxury marinas are being touted when the small population and the access roads hardly justify this. More than twenty-five years on it is still in a sad way.

Portmán comes under the authority of La Unión whose population is little more than a tenth that of Cartagena, which has jurisdiction over the areas around and the quarries that sent their pollutants to the Lavadero that spewed filth into the sea. It is a complicated issue. In general, the people of Portmán don't like the English. They are probably blamed for the high property prices.

While it was good that the contamination stopped, the closure of Lavadero Roberto meant a huge loss of jobs in the mines and their associated offices and laboratories. With the workers being jobless, probably the local cafés and shops suffered a great diminution in trade also. Perhaps some compromise could have been reached with learning how to recycle the waste rather than dump it out at sea. That could have saved jobs and the decades of problems that followed.

While I love nature. I also have great admiration for the ideas and ingenuity of industry. Industrial architecture fascinates me. I would like to see a growth in responsible industry where those that use dangerous substances are responsible for making those products safe afterwards and are capable of disposing of them without leaving terrible contamination in their wake.

Portmán has a rich variety of mining remains. Over the years, I have explored some of the more accessible ones. It also has a couple of architectural gems in the shape of the house of Tío Lobo and the small museum, which was once the miners' hospital. The museum has fantastic Roman mosaics from a villa excavated in the area, peacocks and a woman's face. The museum is so little used the curator is pleased to see visitors.

The house of "Uncle Wolf" nearby is deteriorating rapidly which is sad, as it was designed by Victor Beltrí. It was built for the rich mine owner who preceded Peñarroya in possession of much of the territory. Peñarroya bought the house in time and it became a casino for miners and pensioners. Most town casinos in Spain are closer to cultural centres than what we would understand as a casino. This was around 1950. In 1980, it became a medical centre. It is now empty and ruinous. Some more Roman mosaics are said to be stored inside its walls.

Miguel Zapata was a hard, grabbing boss who was known for setting spies amongst the miners to hear rumours of unrest. Complainants were swiftly sacked. The first doctor of the miners' hospital was José Maestre, his son-in-law. With the family money, he swiftly gave up medicine and went in for politics. Facially, judging by the only picture I have seen of him, he was like the actor, Timothy West.

A huge palace in Plaza de San Francisco in Cartagena, also designed by Victor Beltrí, carries his name. It was the Santander Bank but is now empty. His name is also commemorated in a tunnel, a great piece of engineering that connected the quarries by rail with the infamous Lavadero Roberto.

The carts and engines, mainly German, which were used to run ore from the quarries are left to rot on the hillside outside Túnel José Maestre. A few train aficionados make the pilgrimage to see them. One German train-spotter even wrote to me after seeing a photograph I had posted on the net. It was a diesel train with Sartiaux cars of the type meant to link with Granby tippers.

The tunnel has been unused since 1983 when Peñarroya installed the mill for grinding materials in the San Valentín quarry.

The tunnel itself is flooded.

I have explored nearly two kilometres of it. Water leaks in from a streambed and also from a well en route. The tunnel is built on an incline which is so slight you do not notice it. This, at any rate, keeps the water flowing out. It is very cold and yellow and stains anyone who ventures in. The colour takes at least a day to wear off, however many showers you have afterwards. Fortunately, it is not acidic like the more dangerous water of some of the quarries. It was from this tunnel that we got some of our most interesting specimens. I suspect some cadmium is responsible for the strong yellow colour there.

Five wells connect with the tunnel: Herrera, Descuidado, Nación Española, Carlos y Emilia. The early wells don't seem to add to the flooding problem but one of the later ones does. Carlos is another name for the well, Pozo Mercurio, beneath the San Valentín quarry. It was the site of a mining disaster on the 23rd of October 1970.

The wells were used for loading the train. Six men were with a machine at this point and had the task of filling via the well.

One managed to escape but the others were buried with the machine. It took a couple of days to dig them out and recover the bodies. The bishop of Cartagena visited the quarries and the families of the dead miners. The man who survived decided to go on working as a miner.

A local poet, María Cegarra Salcedo, wrote the following about the disaster.

While her poem was inspired by it, she only speaks of one miner buried alive, and yet, it gives a more vivid picture of what actually happened than the newspaper accounts of the time.

Aquí, el hombre y la tierra
son la misma cosa.
Trágicamente hermanos.
El minero hiere la montaña,
lo maltrata, despedaza, roba.
Un día cualquiera,
el más bello de luz y armonía,
Las piedras, las terreras, dejan su calma,
enfurecen, se dislocan,
se hacen duro mar de ríos macizos,
apretados de agonía.
Aprisionan fuertemente al minero,
enterrándole vivo.
Perdido entre estériles y menas,
permanece muchas horas, días,
haciéndole su gajo, su latido.
Muerto luego.
Acaso en el terrible instante,
el hombre cantaba tiernamente,
pensando en cosas gratas.

Amarga, dolorosa estampa
Que añade una figura más
Al numeroso retablo de olvido.

María Cegarra Salcedo (Poema dedicado a los mineros)

Here, Man and Earth
are both the same,
brothers in tragedy.
The miner wounds the mountain,
tears it apart, abuses it and steals.

One day,
a day of light and harmony,
the stones and earth rise up in anger,
a swollen mass of water breaks its banks,
trapping the miner,
burying him alive.
Lost between sterile waste
and precious ore,
for many hours, or days,
cutting a slash into it with his pulse.
At last, he dies.
Perhaps,
in that terrible last moment,
he sings tenderly,
dwelling on pleasant things.

A sad, bitter image,
that adds another figure to
the tableau of the forgotten.

María Cegarra, from La Unión, was also the first woman in Spain to hold a chemistry degree and opened her own laboratory of mineralogical analysis. She was a friend of Miguel Hernández. She lived to her nineties and the President of the Peña de Antonio Piñana had clear memories of her from his youth as she was a close friend of his mother when he was growing up in Llano del Beal. He described her to me as slightly plump, with a beautiful face. She always dressed elegantly. I feel a kind of empathy with her version, an understanding of the oneness with the earth she describes and also the old battle between the miner and what he is mining. I have read some of her other poems, but this was the finest.

I am almost as fond of crystalline zinc blende with its silvery charcoal facets as I am of amethysts.

The only drawback is the slightly sulphurous metallic smell it leaves on your hands. Otherwise it would be popular for jewellery. It is also known as sphalerite, black jack and false galena. It is at times difficult to tell it from galena which also occurs in the tunnel. The crystals of blende are sometimes combined with quartz or pale amethyst. A dry side passage has many specimens of siderite some of which have tiny beautiful formations of barite on top. The best specimens of blende we found were in areas about half a kilometre inside. After that, a long bare concrete tunnel extends to another section, which has a huge array of stalactites. Their calcite is occasionally crusted with the odd blende crystal. I have only picked up specimens from the floor. I have small ones but have seen some fantastic larger specimens collected by others.

Before entering this tunnel, we spent some months with a series of visits improving the flow of the water that issued through its gates. The area outside is wooded and pleasant for picnics. Our son even found a white truffle in the ground. There are sometimes mushrooms in winter and figs and prickly pears in summer. It is now possible to walk the early stages of the tunnel in wellies. The railway track is mostly sound and you can balance along it, steadying yourself with a pole. Some sections of the tunnel are concrete, others rock. I suspect old mines and caves have been linked together. The tunnel is straight for nearly two kilometres, which makes life easier. You can still get a little daylight from the entrance for at least the first half kilometre or so. After this, the tunnel forks and loops back before forking again. There are old machines at this point, which downloaded material from the San Valentín quarry. They are crusted with calcite.

It has turned red and rusty in places looking like

blood. I did not realise it at the time I visited, but this was the point of the Pozo Mercurio accident. The tunnel to Corta Emilia has deteriorated. I fell down in the water here and therefore didn't do the last section. A friend has explored the entire tunnel up to below the Emilia quarry. He had to wear a full diving suit and said the last lap was the most dangerous and was also devoid of interesting minerals. One day I shall try the last bit out of curiosity.

On one trip we took a friend of my son who had sworn that he had been given permission by his father. We lent him old clothes of my son's as we knew that what you wear down there is irrevocably damaged and comes out patched with yellow. We are in the habit of lining the car with damaged yellow towels for the return journey. We wash some of the worst off in a little stream that trickles out of the bottom of the hill above a drain, stash our wellies in the boot and get home for a shower, or several. Sometimes the clothes worn are thrown away or reserved for future visits. It was a hot day, so son's friend decided to strip down to his underpants for the ride home. We were met at the door by his worried father who was in his eighties, in spite of having such a young son, and had worked as a miner amongst other jobs. He was somewhat horrified to see a semi-naked son, coloured entirely yellow as he hadn't had time to shower any of the evidence away. He had not been given permission and was not allowed out with us again for a while.

Very recently we learned an interesting detail about the tunnel. We were returning from another tunnel mine clad in helmets, grubby clothes and wellies when we ran into a miner's son who was probably close to retirement age himself. He talked of visiting José Maestre tunnel when the trains were running with his father. The tunnel also has a system of pipes running drinking water into the town. The only visible relic connected to this now is a water tank close to the path that leads to the tunnel. Probably the rest went to scrap dealers.

Washing Amethysts in the Bidet

20: *Antiguo lavadero de mineral en Portmán*

17: Portmán

Portmán has now made serious progress with the decontamination. Most of the beach is fenced off and its toxic sands have been mixed with the elements supposed to render it harmless. I recently visited the area to the right as you face out to sea. There is no access there but a member of one of the old mining families showed me a post where he used to moor his boat when he was a child. It is now many metres from the sea. I hope, apart, from the decontamination, they dredge this area so that the sea returns to its old relationship with the land. It occurs to me that Portmán itself, which once had barges taking away the mining spoils or lead ingots, might be a great deal better suited to becoming the macroport than tiny El Gorguel. The road access is a great deal better and it would be taking things back to how they once were rather than destroying a tiny area of nature.

In 1927, a time when boats were regularly shipping minerals to or from Portmán one of Tío Lobo's, the Choncholita, went down in a storm off Carboneras drowning her crew of 8 from Cartagena. Minerals from the Sierra Almagrera, in Almeria, were being shipped on it for processing in Portmán.

With Portmán it is easy to see only the damaged

beach and Lavadero Roberto. But there are many other mining remains. A cold wind whistles through the town in winter. It lacks the sunny aspect of La Unión and seems wrapped in sadness and uncertainty about its future. Other mining remains lie in the hills above. These include Mina Presentación, which has two calcination ovens used for processing manganese on Monte del Pino along the side of the Rambla de Crisoleja. The Orcelitana foundry had a chimney which is visible on the hill above the contaminated beach in an area known as Loma del Engarbo. There is another chimney from Concesión Brandt on the south side. There is also Mina Segundo Ferrocarril with headframe and a tunnel, Mina Amistad with an oven. This is close to the cemetery. Concesión Sivet also had a chimney and ovens. The ovens are close to Lavadero Roberto. Mina Feliz Anuncia had an oven, Mina San Jerónimo, a headframe and Mina San Bartolome also had a headframe. On the west of Cabezo del Pino lies Mina Depositaria, a zinc and lead mine in the Seventies but with older origins. It has the reputation of being a big and difficult mine but is known for some beautiful gypsum specimens found there. It also has some galena and green quartz. It is so labyrinthine that taking string to find your way out again is an absolute necessity.

 The next beach to the main Portmán one is known as Cola de Caballo. It has a small mine called La Gaviota that is simple to visit. It was mined for iron and manganese. There's a variety of minerals there now. What's on the floor is probably better than what is in the walls. There is hemimorphite, barite and gypsum. I have seen some beautiful specimens of the latter excavated many years ago. What's there now doesn't really measure up. The last time we went there, the mine was filled with flies without any obvious signs to explain this like a dead body or turds. I haven't been back since. The beach in front has black sand with glittering particles of chalcopyrite in it. The latter makes this beach unbearably hot on summer days and is definitely not to be walked on with bare feet. Cola de Caballo is on the border between land owned by La Unión and Cartagena. The mine is just within La Unión's boundaries while the land at the top with its remains of a barracks of the Carabineros, belongs to Cartagena.

The hillsides between Cola de Caballo and El Gorguel are rich in barite. Mina Obdulia has crystalline barite. A small entrance now needs a ladder. It is behind a slag heap close to a well-preserved headframe. Mina Obdulia can be reached via a short well-preserved rock tunnel which connects it with Mina Permuta just off the Portmán road. It is on Monte Laberinto. The views from here are spectacular.

A seemingly abandoned house at the entry to Portmán has mining remains alongside. I was told by a member of the Celdrán family that it belongs to his cousin and is still very occasionally used. Some of the hillsides on the way to Portmán were used by the family for hunting rabbits when he was a boy.

A little further inland, just above the cemetery, there are the remains of Mina Segundo Ferrocarril. A long low tunnel leads straight into the hillside. It is slightly flooded. Some good quality galena with a flattened form of crystals is found there. It has become one of my favourite mines for its variety of remains. It also has gypsum, which is partly crystalline and stained with iron, some other bits have small sulphur crystals attached. This appears in the flooded section, which also has an extensive rail system, old pit props and ancient walls. Some of the walls are so ancient they have acquired tiny stalactites.

The best galena is in the upper parts. Unlike most mines, it is one where you go up rather than down. The right branch of the passage runs on after the fork and has a second right turn where you access a small staircase after going up an earth ramp. This leads into an extensive series of passages, which have been marked by string by former collectors. The string leads to a drystone wall that can be climbed to reach the richest galena and further passages and further walls.

Another mine liked by mineralogists is San Timoteo, which is close to the old Roman road and the Monte de las Cenizas. A steep path leads off the main road. There are several openings, which lead to dead ends. Further along the actual entrance is like a small well with a few ledges to take you down a few metres into the mine. San Timoteo is extensive and connects with the Humboldt mine. The Rambla de las Colmenas where it lies has many remains where an old tram or trail track ran alongside spoil heaps. It even has the remains of a garden alongside mining buildings. The valley runs to the foot of Peña del Águila. If you clamber high up among the spoils, there is some exceptionally good barite and other mine entrances.

Fiona Pitt-kethley

18: El Gorguel

We first went to El Gorguel on a trip organised by our mineral club. Halfway along the Portmán road we all pulled over. After some discussion we turned back and took the winding road towards Escombreras. A turning off this led us on to seriously bumpy ground. We wondered if our car would make it and what would happen if it didn't. The road to El Gorguel is little used and no-one can be bothered to remove the rocks or smooth out its hollows. Along the way we passed under a bridge before eventually arriving on a beach rather similar to Portmán but much smaller. There was the same black sand between hills. The entrance to the mine was part way up a hill and invisible from below. Inside there was a maze of passages. We soon found beautiful delicate samples of barite in the walls and started to chip away. This is probably the Santa Bárbara mine.

 We wrapped our samples in torn newspaper and filled a small plastic crate with them. By that time the lights on our helmets and the torches we had taken in, were beginning to fail. Outside, as we made our way down the hillside, I was attacked by a cloud of mosquitoes. It was a warm day, so I stripped to my underwear and went in the sea to cool down.

The black sand was hot beneath my feet. Huge cuttlefish bones lay along the shore attesting to the size of creatures in the sea off there. The beach here suffered some of the same contamination as Portmán. After heavy rain, when the chocolate sand is washed into the sea, old pollution can be seen, compacted grit with yellow sulphurous edges.

My later experiences of El Gorguel came from visiting it alone, turning off the Portmán road and walking down to the beach. The road winds amongst the remains of mines. Few people use it. At the moment, it is scheduled for development. If the powers that be have their way, El Gorguel will become a macro port necessitating the widening of the road. This area was first mined by the Romans. It is possible that the hillside mines will vanish along with the local wildlife if this goes ahead. Much that is charming will be swept away, old mining buildings, hillsides covered with orchids and an old bridge. It is a debatable point whether there is actually any need for an extra container port when the current ones still have some space.

The house I was looking for turned out to be a couple of kilometres inland at the turning that leads down to the beach. Joaquín Toscano, sometimes known as "Toscanini" had the grandiose idea of setting up a rival shrine to Lourdes and his former garden has crosses, rocks and a pulpit as well as bon mots in white paint. It is all rather contrived though. The story admits that he had no followers other than seagulls. The seagulls come from the nearby council waste tip which is now sited unpleasantly close to the straggle of houses.

All Joaquín's poems and bon mots are fading off the rocks. It will soon all be lost. Joaquín's house was supposed to have contained a diabolical image. The exotic garden has lost all but its cypresses.

The mentioned acacia is no longer there. Like the land around the Brunita quarry, it is simply too toxic to support much plant life. His "pulpito popular" where he could stand to recite still exists. He died several years ago but still has family in the area. There are some postings from him left on Facebook, all in capitals, on a group dedicated to saving El Gorguel.

Joaquín was a half-Portuguese mining orphan from Cordoba. His father had been killed by silicosis, a common condition amongst miners. He grew up in the industrial Poblado Repsol and worked for the company. Joaquín used his savings to buy the small houses nearby on the road to El Gorguel and continued to shift large stones with machinery to make his strange rock garden. He believed in the power of rocks and the subterranean and described himself as a gnome although he looks big in photos. He also sometimes referred to himself as "Barba Blanca", White Beard.

In 1992, his house featured in a scene in a film called El Infierno Prometido (The Promised Hell) made by local director, Juan Manuel Chumilla. Joaquín's energy was supposed to have stopped the cameras. My husband seems to do this with chess clocks. At the time, Joaquín had a photo of himself on the bed holding a rosary, and a column of ectoplasm alongside.

The film was on Super 8 but had a more professional remake twenty years later with money from Cinecitta. This spawned another short film on the unmaking of it, a joke on all the films of how things were made. Clips from both are on the internet. I was keen to see the whole thing but felt something on Super 8 was probably no longer viable.

I shut away the idea of seeing *El Infierno Prometido*. It seemed hopeless until I got the programme of the annual Flamenco festival in 2018. The Director was to be given a prize and there would be a one-off showing.

I went to this before opening my mineral stall.

It was quite simply one of the best films I have ever seen. Like Cocteau's masterpiece, it is a take on the Orpheus story, a legend I have always felt close to. It has now been remastered by a process of montage. The film's opening scenes contain a martinete (an unaccompanied Flamenco form to the rhythm of blows on a forge). The Martinete is sung by Curro Piñana. I know his father Antonio, who is one of Spain's top guitarists and is very active in saving local forms and songs.

This Orpheus is a very physical and muscular one. Apart from its visual beauty the film has much to offer. Ginés García Millán is outstanding in the central role. He was to make his name on TV in more conventional parts, but this was his first full-length film. I also liked Franco Citti playing Charon with dignity and Margarita Lozano as Remedios. Remedios is a holy woman who has seen the Virgin. She is a similar character to The Oracle in the Matrix. The extraordinary is hidden within the ordinary. Her dress has three cats' heads in its pattern with goggling eyes. I had done a series of paintings of alien cats and had used these paintings of staring, unusually-coloured cats to decorate the walls behind my stall of minerals at the Cante de las Minas.

The film has a lot of dark humour of the kind that turns up in Alex de la Iglesia and Almodovar. I was a fan of de la Iglesia's earlier darker films and had even worked as an extra on one, *La Chispa de la Vida*. Dark humour and surrealism seem to be part of the Spanish way. I found my own surrealism was unleashed by moving there.

Above all the film is worth seeing for its poetry and use of allegory, Thirst and the search for water are themes, other allegorical things happen with fish and wine. There is a geographical quirk in the Sierra

Minera that makes the setting perfect. The hero, in his journey to look for his lady, jumps over the metal side of the beginning of the Portmán road beside a sign which has La Esperanza crossed out. That part of La Unión is known as La Esperanza. This gives a perfectly Dantesque twist of abandoning hope en route for hell. An earlier scene had been set in the La Esperanza cemetery. The quarry spoils behind once caused an ecological disaster there burying living with dead. While this is not part of the film it is part of the history of the spot. I had used this in a poem and played on the Hope name.

In the trailer to the film, Charon ferries Otelo, his Orpheus, on a rough raft made of a pallet. The quarry is filled with acidic reddish water which is perfect for a river of hell, especially as it leads to the next area which is a ravine known as the Barranco de los Infiernos. I described the washeries by the quarry as a "village of the damned" years before learning about the film. A picture of this area has been a screensaver on my laptop for years. The director angles the camera from the upper sides of the quarry, so the lake looks narrower and more like a river. In the film, a sign, Los Infiernos, marks the area. While this sign no longer exists, the name certainly does on maps. What better spot could exist for the hero's journey beyond hell?

The film stayed with me in dreams. The next night I woke to what I thought was the beat of the martinete.

My neighbours are not into DIY so at 3 a.m. I searched the house for its origin. My husband and the cats were fast asleep. In the end it turned out to be a cockroach tapping the bindings of some old Greek and Latin books on a shelf in our bedroom. I value them enough not to have sold them or given them away, but I have neglected them. My heightened awareness had greatly magnified the regular tap, tap, tap into that of a hammer on an anvil.

I would love to buy a DVD of this film and see it again frequently. I know there is more to find there. There are depths in its texts and its symbolism. Things I missed while simply following the story for the first time. The allegories are beautiful in the manner of Murnau's films, there is the surrealism of Buñuel, and a touch of Pasolini and Fellini. This director's training was partly in Spain and partly in Italy. While Juan Manuel Chumilla-Carbajosa is supremely talented, he is probably not that good at getting his work before the public. That's another kind of skill. A limited edition of a Blu-Ray version, a CD and book in a box was given to the Mayor of La Unión and some of those associated with the film but nothing was on sale to the public. A basic DVD would be wonderful for film buffs. Without that very few people will get to see this marvellous film.

Heaven or hell, El Gorguel lingers on. One of the ecological reasons for keeping it undeveloped is because it is considered the habitat of the trumpeting finch. Those in favour of the port make fun of this bird, calling it an African sparrow, saying it no longer exists there and has not been seen for centuries.

The Sierra de la Fausilla is an area with rare plants and birds. Walking in its sides you can find many orchids in February. There is also a garden-like area of the road with an ancient type of palm. Behind these trees a mine is hidden. I found the rare melanterite in the rubble inside. Most of this will go if development happens. Fortunately, for a while, the EU has put a stop to this. El Gorguel is so close to Portmán that they need to be thought of together in any plan.

Brussels has put a stop to the development, for the moment, as it could interfere with the regeneration of nearby Portmán. The EU has its good points. It sometimes seems to be the only force that can stop the ruin of the environment.

The improvement of Portmán's devastated beach is a necessity that has to be addressed fast. It has already been left far too long. Trying to build a super port next door might be counterproductive. I have found it easy enough to demonstrate about various matters, mostly those linked to ecology or heritage. The police in Cartagena are mostly surprisingly supportive. I have not demonstrated about El Gorguel yet, but there was one instance of confrontation where a lot of people were fined because they did not allow port traffic through while protesting.

El Gorguel has a tiny settlement of beach houses. Most are second homes. Only five people live here according to the census, four men and a woman. For a while, the sea here was used for tuna fishing by the local almadraba method where tuna winds up in a labyrinth of nets and can't find its way out again. The nets are not hauled up until there are enough fishes in the system. This could take weeks but there are frequent checks made to see how things are. Currently there is a large fish farm sited off there. From time to time the nets get broken, an act of sabotage. Sea bass by the thousand make their way out. Many are fished at El Gorguel and Portmán off the beach, others make it to Cartagena and the small beach of San Pedro. I have fished them off the rocks there. It's a bonanza for hard-up locals. What they can't eat is sold on to restaurants or Mercadona. When the fish are let out in El Gorguel some professional trawlers visit and local fishermen put out the odd net and harvest them in tiny boats. Hundreds can be caught in one net overnight.

Occasionally dolphins are seen off this stretch of coast. This again has been the subject of mirth from those who call the trumpeting finch an African sparrow. One politician suggested the dolphins could call elsewhere if they built the super port.

It is likely all of this would be destroyed by any development of the area. Alongside the roughish road that winds off the Portmán road there are also many mining buildings in the Rambla del Avenque. These include Mina San Francisco Javier with its wooden headframe, machine house, calcination oven and chimney, the chimney of Mina Oportunidad, Mina Dios Te Ampare's metal headframe, Mina Laberinto's headframe, Mina Inocente's oven and the newly restored remains of Santa Antonieta. This mine has been improved with the help of a grant which has been used to improve the paths, signpost it and replant some of the areas nearby. Some regeneration of the soil on the old tailing ponds is aimed at. It's a minimal kind of restoration which would be wonderful for the Sierra Minera if ever there was enough money to implement similar ones on other mines.

In the Paraje El Gorguel, there are several more installations which are in poor condition: the Camarón foundry and chimney, Mina Observación a Santelvas with headframe and a powder house, Mina Dos Amigos with headframe and Mina San Rafael with wooden headframe, Mina Concilio-Consuelo only has a tiny part of the machine house and its steam machine left standing. The Santa Isabel foundry still has a tower-like oven, which was used for a unique process. In Spanish this was known as an "horno de viento forzado", literally an oven of forced wind. This type of building is much rarer than the calcination ovens.

Mina San Rafael and Mina La Verdad de un Artista are both in theory mines of El Gorguel but it seemed natural to include them with the chapter on Escombreras as they are more easily accessible from there and are tangled up with its dangerous installations.

Washing Amethysts in the Bidet

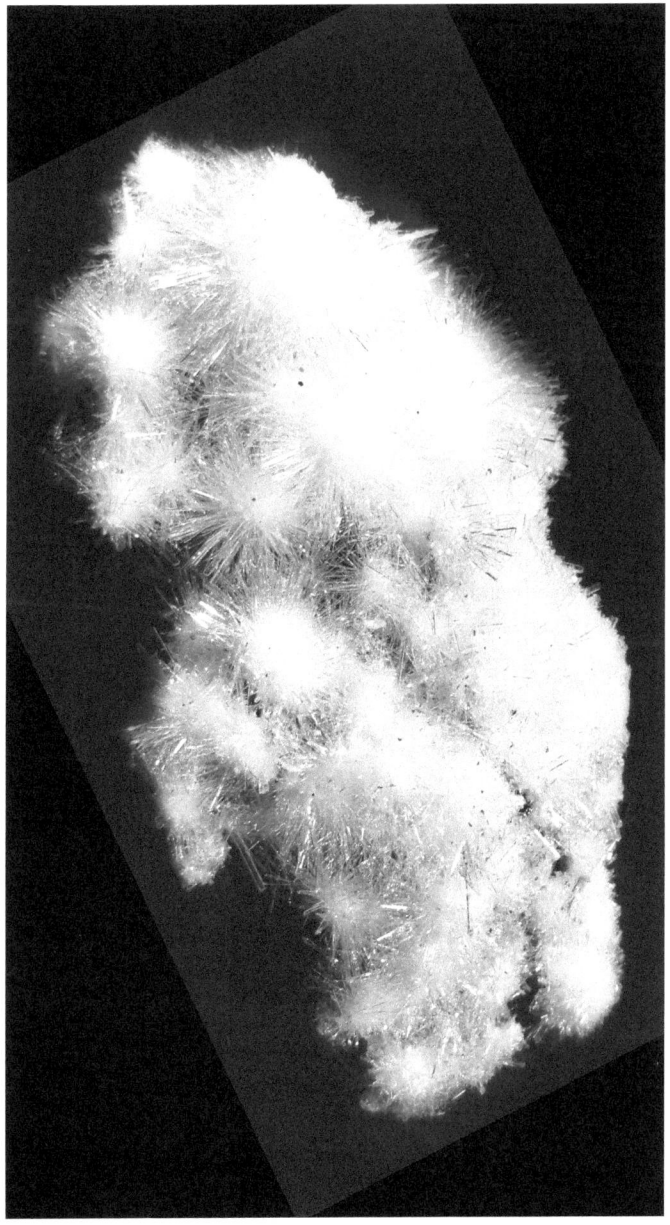

22: Gypsum, Cartagena

19: Gypsum

Spanish mineral names are mostly similar to the English ones. You often only add an a or an o, but gypsum is yeso. Yeso is also the name for one type of plaster applied to walls. Gypsum is one of the most fragile minerals. You can't chuck it into a rucksack and hope for the best. It requires careful wrapping. Even with this it is easy for something that looked spectacular to disintegrate. The largest gypsum crystals were found back in the Seventies in the Iberia mine of Cabezo Rajao on the fourth level at 240 metres deep where it joins with Mina Artesiana. Large crystals were also found in the Sancti Spíritu area. Mina Dificultad, Mina Bragelone, Mina Balsa Depositaria and Mina Segundo Ferrocarril have also yielded good specimens. A lot of the gypsum I have is not from the Sierra Minera.

I have gypsum that is clear as glass from Pilar de Jaravía. Some gypsum of various types exists in the Sierra Minera as well as a lot that is not particularly collectable.

There's a small amount of the glass-like type in the underground quarry near Cala Cortina. It's not as beautiful as that of Pilar de Jaravía though and contains no trapped celestine crystals as that does.

A mine that is popular with mineralogists is Mina

San Timoteo. This is close to Atamaría. It's a large mine inside with access by a ramp of sorts, which is now more like a hole with a few footholds. It looks worse than it is. In reality the footholds work like steps. The entrance is found by going up a small path on the opposite side of the road to Portmán's Calzada Romana. The road winds up into an attractive valley with various ruins. Mina Humboldt is above Mina San Timoteo and was worked with it, although it is not these days considered a good source of minerals. Some of the most delicate specimens of gypsum from the Sierra Minera have come out of San Timoteo, including some that are acicular. Acicular minerals are small, delicate and spiny as thistles. They are usually collected in old egg cartons and shown in transparent boxes seated on cotton wool. To touch them is to damage them. They can only be cleaned with a syringe.

A fire in the mine some years back made areas more difficult to access. Rumour has it that this was started by a collector trying to keep the prices of what had already come from there high. Because of its combination of beauty and delicacy a good gypsum specimen can command a good price at a mineral fair. I managed to get some out intact. The situation is aggravated by further deteriorations in the mine. I hope no collector had a hand in this. One of the beams cut roughly from unshaped tree trunks that holds up tons of rock has come adrift at one end. It definitely would not be a good idea to bump into this. Access to lower areas therefore means crawling low at this point to avoid trouble.

Another beautiful kind of gypsum has been found in the Balsa Depositaria mine. This mine is high on Cabezo del Pino above the Brunita quarry and can be reached by paths hidden in an area along the opposite side of the Portmán road. The entrance is tucked away in the Rambla del Avenque on the other side of the road to where the well lies and several hundred metres away. The entrance is marked by two dry stone walls and a small arch. It has a drystone wall and a short flight of steps.

I have only viewed a small portion of it, having got to the stage where any further progress would have involved inserting myself into one or other hole. I would only be prepared to do this with someone who had done it all before and knew which holes were the right ones. The thought of crawling down a tunnel on my belly does not appeal unless I am sure I can stand up at the end rather than encountering a wall and having to back out again. There is YouTube footage of this mine where those exploring it took seven hours. This is longer than most mineral collectors would contemplate spending in a mine. In my brief exploration, which took less than a couple of hours, I found an ancient miner's shoe rotting on a ledge inside. It is a dusty mine and all the old passages and tunnels are filling up with small debris, making access to other levels more and more difficult The specimens that come out of Mina Balsa Depositaria are mostly long delicate crystals on a matrix. A mineral group from Alicante paid a visit there and got their specimens out intact by hauling them up through the well with a rig they had set up, while the humans emerged through the narrower passages.

There is some gypsum on the land nearby but not of the same quality. It is also being exposed during the enlargement of the San José quarry. It is mostly mixed with other minerals in large flat crystals. Another delicate type of gypsum known as sericulite is found in the La Gaviota mine on the Cola de Caballo beach. Not much of any quality has been found there recently. A way to test whether something is gypsum or not is by running a finger-nail across it. Gypsum is damaged by it where, calcite, for instance is not. This only applies to solider pieces of course, nothing acicular or delicate.

Fiona Pitt-kethley

23: Microscope

20: Quarries, Deviline and Micros

After visiting mines in Almeria and seeing micro minerals I became interested in collecting them locally. Micros are too small to see properly with the naked eye. You carry a jeweller's loupe while looking for them. They are much better viewed at home with a stereoscopic microscope. You can spend hours looking at tiny stones which reveal microcosms complete with caves filled with colourful crystals and geometric forms. It is an addictive hobby.

Quarries are invariably the best spots for micros in the Sierra Minera. Sometimes they only occur in one block of stone in a huge section of quarry wall. Luck and divining talent come into finding them. One of our friends had seen deviline in the Gloria quarries, which lie largely unused at present. I searched them without success, looking for glints of green from copper mineralisations. I was luckier in the San José quarry. It is currently being made ready to receive the waste from Portmán beach. It was only possible to look at it on Sundays and fiestas when the heavy machines weren't rolling.

I found green areas on some rocks. Under a microscope these proved to be malachite which is much more crystalline and less fibrous in character than the rarer deviline. There are many attractive micros here. Under magnification you can see the geometric folds of siderite, tiny, perfectly-formed crystals of quartz, gleaming chalcopyrite, rainbow goethite, etcetera. I kept returning and eventually found some deviline. It was on one large rock only. Once that rock had gone there was no more. I gave some samples to friends from my mineral club. There might never be any more to find.

I thought the heavy machines about San José were solely getting it ready for the residues from the beach. In fact, they were making temporary roads for a curious event. One day as I walked Ruta 33 I saw a huge gathering of JCBs on a flattened area alongside. The machines and their drivers were showing off curious skills to potential buyers. It was one of the most surreal things I have ever witnessed as the JCBs did the mechanical equivalent of line dancing and waltzes. I was the only random spectator. Everyone else had driven there with a view to purchase.

Quarries for building stone can be ancient, like those in Canteras started by the Romans and from which most of the stone in Cartagena came for its theatre, amphitheatre and other buildings. Quarries intended for open-pit mining are a comparatively recent development. In the case of the Sierra Minera they only began in 1953. Corta Emilia was the first to be developed. Between 1965 and 1984, seven more quarries were opened in the following order: San Valentín, Tomasa, San José, Gloria, Brunita, Los Blancos and Sultana. In time Los Blancos had 2 parts known as I and II and Sultana, nearby was sometimes known as Los Blancos III. Because of the methods used the minerals extracted came from a lower level than that known as the manto de los azules, which was mainly silicates. The next level down usually contained pyrite and limonite. While the quarries themselves are comparatively recent most took their names from much older mines in the area.

Huge washeries were built alongside and chemicals were used to separate the metals cheaply. The new processes contaminated the land massively. The destroyed beach at Portmán is the result of this as are the assorted slag heaps throughout the Sierra Minera. Compared to quarries, mines did little damage to the environment. A tunnel bored into a hillside does not interfere with the flora and fauna alongside.

In 1956, a firm was created known as Española del Zinc, S.A. The Fundición of Santa Lucia was also adapted to modern methods. Most of the worst contamination dates from this period.

Cantera Emilia was opened by the Sociedad Minera y Metalúrgica de Peñarroya España. You see the SMMP logo on some old mining installations. Cantera Emilia took its name from a mining concession on Cabezo de Don Juan. It's close to the El Descargador area. The building at the gate of that road is now a shipbuilder's. The quarry was exploited extensively for many years and parts of it are 200 metres deep. It could have been of great interest to mineralogists and easy to develop for tourism as it is close to the road. Sadly, it has just been used as a tip for inert residues. There are now two quarries there in place of the one. It has high security as a lot of old diggers and machines have been dumped there. Some of these are visible if you go on a Sunday, the only time you can look around unobseved. It is possible to wander down from Corta Tomasa. The road is lined with boxes of abandoned core samples. Towards the quarry entrance there are old workshops and other long disused machines in bright primary colours. In spite of their huge size they have a toy-like character because of these colours.

San Valentín is probably the largest of the quarries and the most varied in terms of its minerals. I first became aware of San Valentín from pictures and descriptions on a Dutch site. Two Dutch doctors who were keen mineralogists were discussing its delights. It's in the middle of the Sierra Minera and difficult to access, particularly as its proprietors don't want anyone there. It has been a Mecca for mineralogists because of its huge variety of stones. Apart from our official visit with our mineral club, I took other walks leading me to it and through it on Sundays or fiestas. It's an interesting space. The best-known routes are the long winding road from Lavadero Roberto in Portmán and the turn off from Ruta 33 compete with its fearsome threats of denuncias. There are other possibilities though. A road winds down from the windmills. Another road led out to Llano del Beal ending up near the headframe of Segunda Paz. That has been made impossible now by a cliff fall though. The quarry is 400 metres deep and 1200 metres wide.

Effectively most of the small mines in the areas quarried have vanished. The following list is of those that existed in 1949, on the sites now covered by the Emilia and San Valentín quarries. Amongst these was San Juan Bautista, Maríana and Enrique VIII, which were remarkably rich in lead in the Nineteenth Century, between 15 and 30 per cent. They connected with Mina Belleza. Other lost mines were La Ligera, Bella Unión, San Pedro, San Valentín, Júpiter, Usurpación, La Lucera, Usurpada, Venus, Santa Florentina, Calatrava, Descuidada, Emilia, Virgen de los Ángeles, Tabano, Madrileñita, Felisa, San Jacinto, Juanito, Isabel la Católica, Joséfita, Revolución, Conchita, Nación Española, Descuidado or Descuidada, León Negro, San Isidoro, Afortunada, Frasquita. Apart from Venus, which was an iron mine, these were all lead mines.

San Valentín contains machinery, which was known as Trituración Primaría for grinding up minerals in the first stage of extracting the metal. Rocks were broken down to pieces with a maximum size of 150 millimetres. The remains of the Molienda Semiautógena are also here. It was built in 1983. From there, the pulp was transported via 2.2 kms of tubing to Lavadero Roberto. When Lavadero Roberto II was built in Cantera Tomasa it was taken there instead.

I got in touch with the Dutch doctors whose hobby was collecting stones and picked their brains for a little more info. They had gone via the long Portmán route. They smashed huge boulders for their samples. When, eventually, I paid an official visit with our mineral group, one family did the same and found a large quartz geode in theirs. The rest of us went for smaller stuff.

On that official visit we approached by the road that starts diagonally opposite the El Descargador restaurant and runs up beside the shipbuilder's passing Corta Emilia en route. There is high security, in part because of the parked diggers, close to Corta Emilia. The gatekeeper was a friend of one of our group. His dog was fearsome even by Rottweiler standards and looked as if it could give Cerberus a hard time. It bounced out and slavered for human flesh. Luckily it was on a thirty feet chain and stopped just short of the cars we were in. There was a hasty shutting of windows. Having seen that dog, though I am not particularly nervous of the species in normal circumstances, I would never dream of trying to trespass in these quarries via that particular route.

There's a huge variety of minerals in San Valentín, green opal, monheimite, galena, amethyst, chalcedony, siderite, anglesite, azurite, zinc blende, a rare micro called ECAndrewsite, goethite, cerussite, anglesite.

I wouldn't know what to look for with ECAndrewsite which is named after EC Andrews who first found it in the only other place it occurs, Broken Hill. There seem to be no pictures of this particular mineral on the internet, although it is said to have some similar qualities to ilmenite, a titanium mineral. It probably can only be identified in a laboratory.

In rainy parts of the year there is a small amount of water in the bottom of San Valentín. I believe this is part of the water that supplies the flow in the José Maestre tunnel. The rest probably filters down off the hillsides above. It's not enough to make a lake, like that in Brunita. A rambla alongside the road that leads to the quarry has a tiny stream bed in the bottom which takes some of the water away. I have found a particular breed of brownish praying mantises in that area. Most of the others in the Sierra Minera are green.

Cantera Tomasa was used from 1991 for dumping the waste, which had previously gone into the sea from Lavadero Roberto. Lavadero Roberto II was built there during the last year of mining, It's 200 metres above sea level. Most of its machinery came from the dismantling of Lavadero Roberto. It's a group of buildings that could be restored for mining tourism. Much of its machinery is still in place. The security cameras have been stolen and a door to the main shed is open so you can wander about and photograph all that is left. It is possible to reach it without passing any prohibition notices if you take the path that runs from close to the top of Peña del Águila. It's the ugliest of the quarries and one of the least visited by mineral hunters. A few good specimens have come from there though. On online forums I have seen pictures of green fluorite and light blue celestine. I suspect that these were from the days before the bottom of it was filled with the waste from Lavadero Roberto II. Along the path that leads from it towards Emilia there are many huge boulders, some of which have micros of various minerals, anglesite, for instance.

The smallest quarry was San José, on Ruta 33. Most of what is found there these days is in the form of micro minerals, some of which look very good under a microscope. I have deviline, siderite, malachite, quartz and rainbow goethite from there. Some good smithsonite has also been found. Unlike most other mining areas with malachite there is no azurite alongside. The quarry is now being prepared for the residues from Portmán, so I am glad I liberated some stones while it was still possible. The quarry was started in the Sixties and wound up in 1987. It is not strictly part of Ruta 33 but cyclists usually used it as such as it was easier to negotiate than the pedestrian path nearby, which is cracked and narrow. That is certainly in need of some renovation. The size of the San José quarry is greatly enlarged now with more upper levels. It is possible to walk across those upper levels until you reach San Valentín. There are loads of prohibition notices and warning of landslides, but there is no invigilation on Sundays or festivals. The paths seem sound enough and it is probably only dangerous while the diggers are working there or in heavy rain.

San José and Gloria also had a manto de los azules of greenalite, galena, pyrite, etcetera. San José also had another manto known as the capa negra (black cape) which was rich in zinc in the form of blende and smithsonite, and also had galena and pyrite. A similar layer appears in the Brunita quarry.

The Gloria quarry runs alongside the N345 Portmán road. It was developed in 1965. There are various mines and installations near Gloria, San Francisco Javier, Inocentes and the washery "Buena Suerte".

The quarry is now much smaller than it used to be and less deep. After its closure it was used, with much opposition from the people of La Unión, for dumping waste from the petrochemical industry in Paraje El Fangal in Escombreras. It has the remains of some mining buildings and a flooded tunnel known as Liliane. Some water still flows out from this tunnel and the mud on the floor is bright yellow. It can be waded into but is not good for minerals. There are old mines higher up. Our friend found the micro mineral deviline in this quarry, but in the main, the area is no longer good for collecting. San José and Gloria were wound up in 1987. In 1990, Gloria was used as a place for a group of ETA terrorists to hide out in disused mining buildings. They afterwards hijacked a vehicle and filled it full of explosives, 200 kilos of ammonal. At two in the morning of the 12th September this vehicle was detonated by remote control outside the Guardia barracks in Cartagena, causing 17 people to be injured. The explosion could be heard from twenty kilometres away and also did a lot of damage to the buildings. Gloria is sometimes referred to as two quarries, Gloria Este and Gloria Oeste.

The mines that existed in 1949 on the areas covered by Gloria and San José were: Gloria, Dichosa, Constancía, Encarnación, Santa Teresa Salvadora, San José, Amable, Suerte y Verdad, Buena Suerte..

Corta Brunita was opened in 1984 by the Sociedad Minero Metalúrgica de Peñarroya España (SMMPE). It covers 16.1 hectares. The mines of the area were previously exploited by the Unión Española de Explosivos from the beginning of the Twentieth century until 1950, principally for pyrite. When their factory closed down the area was acquired by the mining boss, Francisco Celdrán Conesa for Minera Celdrán, S.A. This business became Eloy Celdrán, S.A.

in 1953 and continued using the mines and washery until 1980 when they sold to SMMPE. The quarry was only worked until 1988. It has always had problems with flooding. I have never seen it without a toxic lake at its centre.

The mines that were destroyed in the making of Brunita were: Brunita, Los Burros, San Benito, Dicido, El Juanito, Ángelita, La Africana, Lucifer (Satanás), Aries.

Brunita is popular with rock collectors. The Dutch doctors had also paid it a visit. Years ago, it was known for its vivianite and ludlamite but I have not heard of any vivianite being found there in the last ten years. There are interesting ruins along the roadside by Brunita. The first time I visited there I entered from the Portmán road at a point marked by a conglomeration of buildings. These old mining works look slightly scorched and smell of acid. I think of them as a village of the damned. They would make a good setting for a horror film. From there the road winds slowly down into the heart of the quarry. It is littered with chunks of stone that glitter with pyrite and chalcopyrite. Cubic pyrites have also been found there but their size is small. There are some older examples in collections of pyrite on quartz. Some small sceptre like crystals of quartz also turn up in the spoils also. These are very attractive. Dolomite, gypsum, blende and galena can also be found, although I have not seen particularly good specimens from there. Other minerals that can be found are cronstedtite, smithsonite, barite and hematites. On a flat area at the top there are attractive siderite specimens of a golden caramel colour in large greyish white boulders. These rocks are especially hard and each time we have collected siderite we have managed to hammer our own hands as much as the rocks as a result.

From this area you can look down into the Huerta de San Pedro and diagonally across to La Esperanza, which has little these days but a cemetery. If you look at the graves, there were many early deaths of men. Their women joined them much later after visiting the graves with flowers for decades. In 1972, a strange accident with the mining spoils from the Santa Teresa washery the remains of which are beside Brunita buried the living with the dead and wiped out a small row of houses. One man was killed, the caretaker of the cemetery. His body was not located in the mud for several days. Tons of mud descended, and several other people were rescued with the aid of ropes. Bodies were also washed out of graves and the cemetery wall collapsed. There had been massively heavy rains, which provoked the overflow. These days the lake in Brunita is at a very low level. It is a beautiful spot for photographs in a sci-fi kind of way. It is supposed to be highly acidic and also contains a cocktail of heavy metals. This type of mining water has been studied by NASA to see if micro-organisms can survive in it as there are similarities to the conditions on Mars. There are some reeds growing in it on the far side. They seem to have an extremely high tolerance of acidic conditions. The quarry closed in 1988. For a while, a huge Ruston Bucyris machine was left behind. Sadly, this was eventually sold by Portmán Golf for scrap. Another piece of mining history gone.

The Portmán Road was in a slightly different position then. You can walk a small section of it that ends at the foot of Brunita's spoils. On the far side of the quarry you can pick it up again. It runs out quickly there also under the spoils of old mines.

The Los Blancos quarries were opened in 1972 by SMMPE. They covered 26.2 hectares. They ceased production in 1985. They were exploited for sulphur minerals, carbonates from the "manto piritoso" and chlorite from a lower "manto". The mines that were destroyed in its making were: Cocotazos, La Mona, Virgen de las Mercedes, Abundancia, San Sebastián, Vigilante, Rafaela, Jesualda, La Buscada, Tercera Española, Cuarenta, San Carlos, Cuarenta y Cinco, San Nicolás, Monte Carmelo. Several of these mines were acquired in 1952 from the Compañía Minera Bético Manchega.

They are not known for their minerals these days, as too many residues have been tipped there. For a brief period, friends found anglesite there. This had all gone except for tiny poor examples by the time I visited. In earlier times there were various sulphur minerals: pyrite, blende, galena, marcasite, chalcopyrite, arsenopyrite, tetrahedrite, stannite and carbonates including forms of siderite. There was also some low-grade opal, chalcedony and amethyst geodes. Beautiful galena specimens were also found there decades ago and some chrysocola of which there is a specimen in the La Unión museum. This is all rather a loss from the mineralogist's point of view. Another area from the "mantos" was known as the "capa negra" (black cape). This contained marcasite, pyrite, blende and galena. This was also present in the quarries of San José and Gloria. There is a small greenish lake in Los Blancos I, which does not look as acidic as that in Brunita. The current form of the quarry is due to the tipping of residues from Balsa Jenny in 2003. Some work goes on there from time to time, diggers flattening paths, etcetera. I am not sure why and have seen no public plans.

While I was trespassing on the edge of one of the Corta Emilia quarries, I saw an old signpost for Los Blancos. Probably this was for Los Blancos III, otherwise known as Corta Sultana. It is the nearest of the three, though not really connected by today's standards.

The Sultana quarry dates from 1984. It covers 24.8 hectares. It's so close to Llano del Beal that you can hear echoes of the voices of those who are outside talking in the street. The quarry contains the remains of a vaste Ruston Bucyris digger for those who like to look at such giant industrial toys. There is also a unit from one of the old aerial cables. I took my son's photograph sitting inside that. These cables once threaded the Sierra Minera. Here and there you see the remains on the ground. The mine in Corta Sultana at one time had remnants of straw in it from when horses were stabled there. It is still visitable though it is easy to miss the entrance which is tucked away behind a heap of spoils.

From the top of this heap you can clamber down. There have been so many rockfalls due to water that many of the passages are now blocked. The entrance must have been a great deal more accessible in the days when horses were used. We got the impression that we could still smell the horses in one area of the mine though any remnants of straw had gone. A flight of very broken steps leads to a lower level, which has broken wooden ladders that were used by miners to reach the lowest parts. This whole area is beginning to look rather dangerous with a long crack in the cement archway above the stairs. There are tiny stalactites and incrustations of calcite and goethite on the walls of the few parts that are still visitable.

Every year the digger gets rustier and loses more parts. It's a useful landmark when you are on the hills above, as one quarry can look much like another. Only one of the local ones has a digger.

The quarries are a lunar landscape with comparatively little vegetation. Some have acidic lakes, reddish in tone, where the pyrite in the soil has mixed with rainwater to form a weak solution of sulphuric acid. Nothing grows near these.

Replanting has had an almost total failure rate.

The only things that can tolerate such contamination are reeds and tumbleweed. The introduction of quarries changed the landscape hugely, lowering the top of Sancti Spíritu in the case of San Valentín. The Sierra Minera was once covered in forests. Mining and local ship-building took their toll. Several centuries ago, a huge amount of timber went into the making of the Armada. In the following centuries a great deal was used for smelting. There was no coal in the area, so ovens were stoked with local wood.

Perhaps even more than this, the quarries with their huge open spaces took away the forestation. There are occasional attempts to put it back. All but about ten per cent of new trees planted here and in the other mountains of Cartagena die off. Usually it is lack of aftercare and planting at the wrong time of year. I have taken part in some replanting programmes and hope the new ones succeed better.

Time will tell.

The policy now is to plant only in winter. Each baby tree gets a small stick and some mesh to go around it. You dig a deep hole and clear stones. The plant is sited in the middle with a kind of raised ridge around the space to make a well for moisture. If roads are close enough to planting areas, the fire brigade waters the mountainside when everyone has finished. There is no shortage of people willing to volunteer. Usually hundreds turn up and make it a family occasion with children alongside. A lot of Cartagena cypresses are used these days. They are less vulnerable to pests like the processionary caterpillar, which is killing off many pines and is also a hazard for humans and dogs. If you are unlucky enough to touch one its toxins raise a huge rash on your hand, including blisters. I have had this happen occasionally.

The Cartagena cypress, tetraclinis articulata, is a rare tree that was dying out and it is also less easily damaged than pines, which go up like a torch in a forest fire. There is a LIFE project, dedicated to replanting it in the area. I went to a conference they gave in La Unión. Saw palmetto also has some ability to survive. The top burns off, but the roots remain strong and put out new growth soon after a fire. Other trees and shrubs that might be planted by various organisations involved, such as ANSE, CreeCT and ARBA, are buckthorn, tamarisk, arbutus and wild olives. The fig trees seem to manage to plant themselves.

24: *Maquinaria del Lavadero Roberto en Portmán*

Fiona Pitt-kethley

21: Lavadero Roberto and Peñarroya

I had explored Túnel José Maestre many times, attracted by its minerals. It was years though before I had a good look over Lavadero Roberto. Lavadero means washery. It was built in 1957 by Peñarroya. Open pit mining had become the favoured method by then. The works included three grinding mills and various flotation systems. It was capable of treating 2400 tons of rough minerals a day. The first load of minerals came through the tunnel on the electric train in July 1957. In 1965 it was connected to the Gloria quarry also. I assume this was by road alone as the Gloria quarries are not that close to Emilia. In 1966, it was connected to San Valentín, which was easier as that is en route for Emilia and a little closer. This was at about two kilometres along the tunnel where there is a loop and a division in the passage. A rusted piece of machinery lies at this point which looks like part of a system to bring down minerals from a pit above. The pit in San Valentín is now closed. That is as far as I have explored in the tunnel.

Lavadero Roberto was improved from time to time and considerably amplified in 1972.

At this period Peñarroya was extracting more than 2.6 million tons of minerals a year from the Sierra

Minera. It was, by the time it was closed, the biggest in Europe in terms of the amount of minerals it could treat.

The plant treated the minerals of the Manto de los Azules, the Manto Piritoso (rich in pyrites), and the different mineralisations of Brunita and Sultana.

On the 31st March in 1990, the environmental authority ordered a stop to the dumping of waste in the sea. For thirty-three years, the Lavadero had been putting 7000 tons of waste directly into the sea, a cocktail of mining waste, chemical products, earth and water. Undoubtedly the chemical products were the most harmful. In all, 50 million tons of sterile waste are reckoned to have been deposited, affecting eight square kilometres along the coast to a depth of 150 metres. The chemicals used in flotation were cyanide, xanthates, copper sulphate, caustic soda, sulphuric acid and zinc sulphate. Xanthates are the least well-known on this list and one of the least toxic, but even they can cause problems to aquatic organisms. Not an ideal addition to the sea, therefore. Zinc Sulphate has been used in animal feeds but is toxic in excess. It also has been used to stop moss growing so probably not good for algae in the sea. The rest of the list are so obviously toxic and dangerous and need no further explanation.

For a while, the Lavadero remained open, and parts of it, including a great deal of machinery, were used to construct Lavadero Roberto II, next to the Tomasa quarry. This building was similar in shape and is more intact than the old one. Its construction was so close to the end of mining in the area that it was not used for long.

Initially, Lavadero Roberto was easier to explore than its successor. It is open to the winds. The easiest approach is along the railway line that connects it with

Túnel José Maestre. There is also a huge fire escape at the front. When I first passed this area, this was open to those wishing to make the ascent. It is now fenced off, as is the start of the road to San Valentín from there. But the fence is breached periodically by local vandals. There are odd missing steps on the stairs now through rust, so care is needed. The steps are concrete with a metal surround and metal rails. The inside of the building gives the feeling that it was left at a moment's notice. This is true of many of the local relics of industrial architecture, which adds to their mystery. Was it indecision about the future of mining that caused this? Surely it would have made sense to sell machinery and the train at the close down of mining. Perhaps attempts were made, and no buyers found at a stage when many other mining enterprises were also closing down in Spain and elsewhere.

The interior of Roberto has several levels. On my first visit it was possible to enter the top one. A few years later, a section of walkway which was previously merely precarious, had fallen. On the first visit my son and friend were longing to enter a tunnel leading off Lavadero Roberto. I explained this wasn't a good idea as it would probably still be severely contaminated and also, it would not contain any minerals to be worth the risk. The lower sections of the interior contain the remains of conveyor belts and grinding machines and other objects so badly decayed it is hard to know what they were used for. Metal rusts away given time, leaving other parts like rubber, which takes longer to pass into nothingness.

In the early stages after it closed Lavadero Roberto could perhaps have made a good museum for local mining history. Sadly, it is now just a monumental relic, visible from a distance in its sinister decay. Recently some men were arrested stealing metal from its interior.

Yet more mining history lost. It is still relatively easy to get in via windows abutting the staircase. I have done so to take photographs.

After speaking to the miner's son we met close to Mina Segundo Ferrocarril, I began to see the other side of the closedown of the contamination of Portmán. Portmán, doubly screwed, first by having its fishing industry wiped out with the filth pouring into the sea and secondly, by the total closedown of all mining. A village with nothing left and almost everyone out of work. Portmán Golf is not exactly in Portmán and workers who have some English are bussed in to act as waiters, etcetera. It's rumoured to be a good job with tips. But they are bussed from La Unión, which does little to improve the dire work situation on the spot.

What was Peñarroya really like? What did working for it feel like? It had started as a French company but gained a Spanish name after the Rothschilds left. A few of its logos, SMMPE can still be seen on buildings across the Sierra Minera.

When I was keeping a stall at a Collectors' Fair in the Salinas of San Pedro, all the stall keepers dined together at an excellent restaurant alongside. Manuel Morales was there, also selling minerals and mining memorabilia. We chatted a bit about my researches at lunch and he was kind enough to give me a booklet produced by Peñarroya while it was still a huge power in the Sierra Minera. Here were all the figures and methods of working. Ironically, the book's last chapter was on ecology. Even at the height of the contamination they were experimenting with plants on the beach at Portmán. I was reminded of Repsol's more recent attempts to be seen as a force for good in terms of replanting.

I have always found the role of Devil's advocate a tempting one so I will quote some of the facts and figures from this little book, produced in 1985. It talks of the area owned as the "Centro Minero de la Sociedad Minera y Metalurgica de Peñarroya-España, S.A: en Cartagena La Unión. It was 10 x 5 kilometres of the Sierra de Cartagena. The main processing works was in Portmán alongside the sea.

The booklet opens with a mention of the area having been mined for more than 2000 years. It describes the modern mining of the end of the Nineteenth Century as anarchic and funded in a small way. Anarchic, presumably because it wasn't in the hands of just one big boss. I suspect that the anarchic period was one of the only ones where workers were not radically exploited. If you owned the concession, you could at least decide on the hours worked. Production went up hugely with the opening of quarries from 1957 onwards. In the lists, San Valentín and Tomasa are lumped together. The only quarry not listed in this book is the small San José one. Perhaps it was considered part of Gloria at this stage.

Ore from lead, zinc and iron was processed in Lavadero Roberto. The resulting blende, galena and pyrite concentrates were sold in Cartagena. The galena went to the Fundición Santa Lucia. The blende went to La Española de Zinc S.A. and the sulphuric acid plant of Explosivos Rio Tinto S.A. which was later known Potasas y Derivadas, S.L: All that is left of that now is a polluted wasteland on the outskirts of the city. The pyrite went to three factories in Cartagena. Peñarroya was only interested in treating the richer ores. Poorer ones were sold to Minera Celdrán S.A.

At the stage when the modern quarries were excavated, exploitation had become far more scientific. There is a sense that older mining involved more chance and instinct, something akin to ancient water-divining. A mineral collector would of course be good at this as they would know by the stones in the hillside what metals existed in a particular area.

Peñarroya seems to have done more by way of core samples and sounding to establish where the best minerals might be. The employees I have met who worked there in its closing years had science or engineering qualifications and were concerned with research in the Sierra Minera and further afield just before the great closedown of mining.

There was analysis of the blocks prepared for processing. Certain sulphates and bismuth with galena could damage the contents in a flotation tank. This stage was known as "preexplotación" or "control de leyes". Inventories were made of the reserves waiting for processing. And the material from new areas being worked was assessed for its potential. All was part of a careful plan for the future working of each quarry, which extended six years into the future.

The small subterranean workings were converted into open pits. In reality, what Peñarroya saw as anarchic, had been minimal in terms of pollution while opencast mining created deserts. It is sad that no necessity for control of contamination was perceived as necessary at this stage.

In the concentration process at Lavadero Roberto, seawater was used because of the lack of an abundant supply of fresh water in the area. At the height of its operations the consumption of seawater was 800 cubic metres an hour. The use of seawater created a situation where regular controls on materials and equipment were needed because of the greater problems of corrosion. The water was pumped from the sea at a point below the lighthouse. The initial breaking down of rocks was done in the semi-autogenous mill at the San Valentín quarry. The pulp of rocks and water descended through large pipes with the help of gravity to the concentrator at Lavadero Roberto. Many of these pipes still exist and can be seen by a high path on the rambla between Lavadero Roberto and San Valentín. Part of the area can be walked. The pipes run out and are in their most broken state close to the works that are outside the quarry. I have put my camera to a small hole in the large closed metal building here and a flash photo reveals that much of the old machinery is still in place. Once at Lavadero Roberto the pulp was milled again to break the particles down to a smaller size. There were then four phases of flotation. The first three separated out blende and galena, while the fourth separated pyrite from the mix. The blende and galena went through further milling and separation from each other. Once all useful metal ores were separated out, the waste went into the sea.

Apart from the mills and flotation tanks, etcetera, there was considerable office and laboratory space with computers galore. Computers were rather large in bygone days. There were 30 workers in mineralogy and analysis. This part of the business was known as CEAM, Centro de Estudios y Analisis Mineralurgicos. This centre was open 24 hours a day and could also coordinate help when there was an emergency. Most of the work related to the ongoing processes in the Sierra Minera, but some research was also undertaken for third parties.

Eventually Peñarroya merged with the Germany company Preussag, A.G. to become Metaleurop, S.A., which in turn became Recyclex, S.A. which still exists though Peñarroya is no more.

There is an interesting and sad detail in a contemporary article in El Pais on the closing of Peñarroya. The staff tried to give blood but found there was too much lead in it. They had sacrificed their health and now their jobs had gone too.

Washing Amethysts in the Bidet

25: *Gran Hotel*

Fiona Pitt-kethley

22: The Big Bosses, Strikes and Riots

There is a small genre of adventure fiction that concerns mining. It overlaps with treasure literature but has a slightly different character. The bulk of the books in this genre are American and Canadian. I have read various samples. After a variety of tribulations, the hero finds the ore and the mine pays out. He usually pulls the heroine also. The mines are always gold, silver or copper. Anything else would be lacking in romance. Lead and zinc never feature in these stories. The romance of the treasure-hunting element is offset by realistic pictures of shanty towns and the seediness of mining life. What this branch of literature usually does not portray in detail is the spectacular difference in lifestyle and life expectancy between the miners and their bosses. Some of the big bosses in the Sierra Minera made it up from being a foundry man or a shopkeeper but not from being the most basic labourers in the mines.

The reality of mining in the Nineteenth Century and early Twentieth was that it was spectacularly exploitative with horrendously long hours and terrible conditions. Perhaps worst of all, there was no way the salary earned could cover basic needs. The only hope of getting by was to have several earners living within the same small hovel to cover the rent, etcetera.

By contrast, the mine owners lived in huge palaces and mixed with the cream of society. Most of the palaces they commissioned still exist in Cartagena and La Unión. Most are now owned by banks. Even in death, the mining bosses were spectacularly richer than their employees.

There are architectural gems by way of family pantheons in the cemeteries of Santa Lucia and La Esperanza. You can't take it with you but you can be buried in style leaving a monument for your family to visit, which will proclaim your name long after the mines have closed.

There is a romance about the mining industry that leads people to see it as a treasure hunt in which a humble man might get lucky. It is, in fact, something rather different. A history-related forum I belonged to posted a map of the location of a copper mine that was based in Algameca, an area on the other side of Cartagena from the Sierra Minera. Someone suggested that the location could make people rich. Knowing a bit about Algameca I figured out exactly where the mine used to be. I quoted a Spanish rhyme by way of reply; "Mina de cobre, minero pobre". (Copper mine, poor miner.) A mine is an opportunity that has to be financed and handled right. It is not a pot of gold at the end of the rainbow. And an abandoned mine has often been abandoned for a reason. It couldn't be made to pay. It was no longer worth the effort.

The mining bosses were clever men who learned to make technology pay. Some were also cruel and exploitative in the extreme. The benign bosses had a lifestyle far different from their men but did at least put a bit back into the system by initiating things like poor schools, hospitals, pay outs for the injured. The others simply worked their men into the grave while taking everything they could from them via shops they were forced to buy from. Any complaints about conditions were met with immediate sackings.

One of the earlier names connected to the area's mining is that of Hilarión Roux, a banker from Marseilles, connected to the Casa de Rothschild, the French arm of the Rothschild family. Hilarión Roux, was born in 1819, and went to Spain as the representative of the Casa de Rothschild in 1842. Though young he had experience of the world of metals and lead in particular. From Alicante he went to Cartagena. Silver was becoming important at this period and he saw possibilities in the foundries of the area, such as the Franco-Española in Santa Lucia and that of San Jorge in Escombreras. In 1844, he married María Piedad Aguirre Aldayturriaga. Her brothers, Eduardo and Simón were involved in industry. In 1875, Alfonso XIII gave him the title of Marqués of Escombreras. By 1877, he had acquired, or had an interest in a huge number of mining concessions. These included Cuatro Santos, Braguelona, Eloisa (El Corcho), Pronta, Virgen del Carmen, Júpiter and its annexes, Porvenir or Anticipada, Lucera, Calatrava, El Ángel de la Guarda, Santa Teresa, Salvadora, Dichosa, Emilia, Catón, Constancia de un amigo, Colmenera, Encontrada, Grandeza, San José, León Negro, Crescencia Segunda, María Dolores, La Loca del Capellán, Rómulo, San Rafael, Reserva, Reforma, Virgen del Pilar, P. Hilarión, Zurbano, Pobrecita, Amigos consecuentes, Pablo y Virginia, Previsión, Santa Filomena, Santa Ana y San Juan, Observación, Nunca Vista, Isabel la Católica, Iberia, Oriolana, Torremendo, Violeta, San Simón, San Joaquín, San José, La Chapinas, No te escaparás, Sin duda, Por si acaso, Torrente, El Tábano, Revolución, San Joaquín, Virgen de los Ángeles, Perdida, San Eloy, San Manuel,

Si puede ser, Vigilante, La Paloma, San Aniceto, Cometa Donati, El Español, Edetana, Inocente, Plutón, Precaución, La Suerte, Vulcano, San Bruno, Ebraldo, Africana Segunda, Olvidada, Marinera, Las Cenizas, Herculano, Pompey, Saint Ginés, Inocente, Emma. He also acquired Isabelita, Virgen de las Mercedes and Cocotazos, which three were in nearby Algameca rather than the Sierra Minera. He also owned others in Mazarrón, Águilas and Lorca. All these were transferred to the Compagnie Française Des. Mines et Usines d'Escombrera. The concessions state their areas which are slightly different to the modern town divisions. The top of Sancti Spíritu belonged to the Garbanzal part of La Unión and El Gorguel was part of Alumbres. Portmán is sometimes written as Portmaur. Some of the names of mines are very familiar, others have disappeared completely.

Another name with a Marseilles connection that appears around the same time as Hilarión Roux's is that of Ignacio Figueroa y Mendieta, Marqués de Villamejor. He was of Spanish descent but was born in Marseilles. The large lead foundry there led both men to have an interest in its production. Figueroa also went to Spain and had mining interests in several areas, including Cartagena. From mining he went into politics. He was a cultured man who spoke several languages, wrote and painted. In 1884, he ordered improvements to the port at Portmán so that iron and manganese could be exported.

The other personalities that live on from mining are in many, but not all, cases connected to palaces and grand houses in Cartagena or La Unión. They are Esteban Llagostera, Joaquín Peñalver Nieto, Eloy Celdrán, Andrés Pedreño Torralba, Miguel Zapata Sáez and his son-inlaw, José Maestre, Pedro Conesa, Bartolomé Spottorno, Serafín Cervantes, Camilo

Aguirre, Francisco Dorda, Pió Wandosell, Celestino Martínez. Recently, I became friends with the nephew of the mine-owning Celdrán. He is the President of a Flamenco club I belong to. He has some interesting tales from his youth. His parents had beautiful mineral samples to show in their house. The Celdrán company owned many mines and washeries. Apart from those at El Lirio, they also owned Julio César, Rosa, Jenny, Segundo Paz and the Los Pajaritos group. All of these are close to Llano del Beal.

The Romans used slaves to do their mining. The Nineteenth Century bosses were only marginally better.

Towards the end of the century the anarchist movement grew as a result and strikes and riots disrupted the mining world. In the early Twentieth Century, the socialist movement took over and gradual improvements were made. In the Sierra Minera, the new laws were often ignored, and people did what they had to do to keep their jobs. Sometimes shorter legal hours were instituted, but usually only if the worker agreed to accept a lower than normal salary. Pay for men was usually twice that for women and three times that for children. This meant that bosses were happy to employ a great many children.

In the Sierra Minera, it is probable that lead with a small side order of silver was the biggest fortune maker. It certainly was in the case of Pió Wandossell. I became interested in this colourful figure after using his former foundry as a picnic spot on various occasions. A book on him by a member of his huge family gives the details. He was not so much a prospector as a skilled foundryman who rose through the basic skills he had learned in his hometown of Alhama la Seca in Almeria. He was not only a mining empresario but also owned part of the bank in Cartagena, a dry dock, an electricity company in Orihuela, dabbled in politics and was a freemason. Three of his twenty-four children were in the first football team that became Real Madrid. My impression is of a fairly urbane industrialist who had charitable concerns also. His portraits present the appearance of a dandified Edwardian with a huge moustache. He also owned an immense amount of property in the area and in other parts of Spain. Here are some of the important dates from his life drawn from the book written by his great grandson. He was born in 1847. In 1867 he founded the society "La Amistad" and became the owner of the Talía mine in Mazarrón. In 1868 he moved to La Unión to the house of a relative, Manuel Rodríguez Gil. He rented an oven in the La Paz foundry in La Unión. In 1870, he married his first wife, Dolores Calvache Yáñez from his hometown. What is now known as Huerto Pío was then called Villa Dolores after this wife. In 1871, he leased the mine Esperanza. In 1872, he rented the Tres Hermanas Foundry in la Unión. In 1873, he bought land for his first house. 1n 1876, he sub-leased four mines in La Crisoleja in the Sierra de San Ginés. In 1878, he bought land to build the "Dos Hermanos" foundry in the El Descargador area (our picnic spot). In 1879, he founded the mining society "Wandosell y Toledano". In 1880, he founded the society "La Familia" with Manuel Rodríguez Wandosell. In 1888, his wife died after bearing him thirteen children. Several of these children had died also. He married Francisca Calvache Yañez, her sister, a few months later. She bore him a further eleven children and outlasted her husband by sixteen years. In 1891, he became a councillor in La Unión and also had "La Caridad Hospitalaria" built. In 1892, he bought Mina Oportunidad in Portmán with Ignacio Góngora. He passed over management of the Dos Hermanos foundry to his son, José, at this stage. He bought the land also where Villa Dolores was built. He presented an offer to build a dry dock in the Arsenal in Cartagena and became part of the management of a team proposing a project of a tunnel between La Unión and Portmán. 1896, the dry dock project started. He did a shared lease with Miguel Zapata for the mines Joaquína and Victoria. He passed control of various mining businesses to his friend, Andrés Teulón, another freemason. In 1897, he sold the Sociedad Franco Española de Explosivos in Alumbres. In 1900, he became part of the founding of the Banco de Cartagena and the Compañía Cartagenera de Navegación. 1909, he bought the lavaderos Anita and Los Quemados in Llano del Beal. From 1910, his health began to deteriorate after a stroke, and more and more was delegated to either sons or friends. In 1916, when the miners were shot outside the Dos Hermanos foundry he was in Madrid at the time and had nothing to do with this. He died in 1920, at his house in Calle Carmen, Cartagena.

Soon after moving to Cartagena I had copied a colourful article about Pió Wandosell off an online magazine. I saw that my husband was about to play someone surnamed Wandossell in his next chess tournament in Murcia. I thought I would pass a copy of this article on. When we got to the tournament there was no Wandossell to be seen. Several games were played when eventually he turned up about halfway through the tournament. I lost my nerve as the current Wandossell generation turned out to be a kid of ten or so. He was in football gear and had arrived late because of training elsewhere. With a rare surname like that plus the football enthusiasm it is pretty certain he was a descendant.

Pió Wandosell was, on the whole, a good boss. Others, were less popular. A less well-known name amongst the mining empresarios was that of Celestino Bonifacio Martínez. He was born in 1858, in the village of El Estrecho de San Ginés to a humble family. He spent 28 years as a shopkeeper before buying into mining in the Sierra Minera, Badajoz and Ciudad Real. In 1891, he won the contract for street lighting in La Unión. From 1894 to 1897 he was a collector of local taxes which did not help his popularity. He had an attractive home in the Plaza de la Merced. It has a window like a wheel at the top. It was designed by the architect, Tomas Rico. Martínez was an unpopular boss, who was seen as the face of oppression at the time. His house was destroyed and looted in the General Strike and riots of 1898. I am presuming that this was another house as the one in Plaza de la Merced is still standing, although a little rundown and split into separate flats. He had to flee to escape lynching by the mob. The rioters went on to La Esperanza, Cabezo Rajao and El Algar after leaving much damage in Cartagena. The Mayor of La Unión was captured and injured. A regiment was brought from Seville to help quell the riots. In 1900, with several other friends, he founded the Compañía Cartegenera de Navegación. Its four boats were named after the four saints of Cartagena and exported local produce. Celestino Martínez's wealth was put amongst other things into the Gran Hotel designed by Tomas Rico and finished by Victor Beltrí. The idea came for this, when his family was having difficulty finding accommodation in the area, one rainy night. He realised that there was a place for an upmarket hotel. Like Wandossell, he left many descendants. Recently a book about the Gran Hotel and Celestino Martínez has been published and the centenary of the building was celebrated with a small re-enactment in the square outside and on the balconies. The Mayor of 2016, José López Martínez, played the Mayor of the time. The band of the Infantería de Marina played pasadobles outside. The building is now a kind of commercial centre with offices above a couple of banks.

Miguel Zapata was an unpopular boss. He was born in 1841 in San Javier from a family in farming. He was known as Tío Lobo after he killed a wolf, which was attacking their cattle. The stuffed wolf's head was displayed in his house in Portmán, eventually, and had a light bulb installed in its jaws in later times. From San Javier he moved to Llano del Beal where he kept a small liquor shop. (There are Zapatas, possibly of the same family, still in the wine business in La Unión.) He soon started to invest in mining operations. He was involved in the foundries of La Orcelitana and La Purisima Concepción. His daughter, María Visitación Zapata, married José Maestre Pérez, who was initially the doctor in Portmán but soon went into politics. When she died, José Maestre married another daughter, María Obdulia. José Maestre was in turn Mayor of La Unión, diputado for Murcia, minister for a short time of Supplies and later of Transport in two governments of Antonio Maura, in the reign of Alfonso XIII. In 1921, he was also the governor of the Bank of Spain. A local artist's model and prostitute, Caridad La Negra was his mistress. She became famous later as one of those who managed to save the Caridad church at the beginning of the Civil War.

In 1890, Miguel Zapata founded the Maquinistas de Levante. Its ruined buildings are close to the FEVE station in La Unión. One part of them is used to store old mining relics such as abandoned machinery and waggons. It was initially used for making mining machinery of all kinds for the mines of Cartagena and Mazarrón. Another office from that complex is used for the annual meetings of my mineral club. In 1997, an attempt was made to revive ithe Maquinistas de

Levante with the German company KSK. All the local dignitaries turned up for the inaugural act. A hundred workers were to be employed. Initially there was a contract for 1200 drilling machines. Far less drills existed, and some creative accounting went on and the whole project came to grief. It was Germany's biggest white-collar crime.

After visits to Cardiff and Swansea in 1874, Miguel Zapata installed aerial systems for the transport of minerals. The first was between the La Crisoleja mines and the docks in Portmán, where minerals were treated, and residues dumped in the sea. In those days, the bay of Portmán was entirely different from the contaminated waste it has become, and container ships could be loaded with minerals there. The cable system was 2 kilometres long with a shift of levels of 158 metres. Here and there in the Sierra Minera you can still stumble on old cables and posts and parts of the aerial systems. Ten years later he installed one at Mina Lucera of 2260 metres with a displacement of level of 268 metres. This was designed to connect with the railway to transport it on to Madrid, Alicante or Zaragoza.

In 1913, the beautiful house built for him by Victor Beltrí was known as "La Casa Grande". It is completely ruinous today but shows signs of its former grace. He also founded a hospital for orphans of miners in La Unión. At the beginning of the Twentieth Century, he started importing lead for processing from Tunisia, Algeria and Morocco. One of his boats was the Malabar, a British vessel. Steamers plied their trade between Portmán and Cardiff and Swansea. The stretch between Swansea and Port Talbot has many huge metalworks. My grandfather worked in one before becoming a preacher. The boats returned to Portmán filled with Welsh coal. It is hard now to visualise Portmán as an important port. Photographs of it at this stage seem to be rare. It must have been not unlike the highly industrialised Escombreras in those days. In 1918, Tío Lobo died in his house in Portmán and his mining empire was left to his son-in-law, José Maestre. Tío Lobo's reputation during his life was not a good one. He was known for planting spies among the miners to check for potential unrest or industrial action.

One of the earliest short films made in the region (1900) was "Salida de Operarios de Don Miguel Zapata en La Unión". Mining conditions were harsh. In the last quarter of the Nineteenth Century, twelve hour working days or nights were commonplace for both children and adults. It wasn't until 1900, that an 11-hour working day was established for fourteen-year-olds. While there are no children at work in Spain these days, both agriculture and the film industry seem to operate very long days and get away with it. I worked twelve-hour nights on a film for a pittance several years ago.

Agricultural work in some cases is paid by what you can pick with no pay for rainy days. I know people who are working as hard as Nineteenth Century miners at that, but at least they are out in the fresh air. In the Nineteenth Century and early Twentieth, the youngest workers would run the galleries with 20 kilos or so of rocks per load. I have sometimes carried this much on hillsides but have never attempted to run with it. Typically, a young worker would run 100 metres or so with a load before dropping it off. About 90 or so of such trips would be made across the day. Adults, in many cases were working in cramped conditions where they had to stoop all day. I still see many low passages without much headroom in the mines. Miners were mostly short.

Some mines were also flooded so they worked in water. The lowest levels might also be short of oxygen.

Those who worked on the surface were not necessarily better off. The molten metal of the furnaces and ovens gave off noxious fumes. Many mines and the factories that processed metal were open all day and night with workers taking shifts. For a short period, women worked in the washeries and factories but from 1878, their jobs went to children and teenagers. Some of the larger mining businesses provided a tiny amount of help for widows and families after accidents. Many didn't. In 1910, the law reduced the working day to 9 hours inside a mine or 10 hours outside. This law was largely ignored in the Sierra Minera.

Other compensation laws were introduced but they were often got around by laying blame on the worker. The first two decades of the Twentieth Century were probably even worse than the years before for workers. What was paid was not a living wage. A soup kitchen helped some families with standard fare of a stew of green beans, potatoes and a touch of fatty pork. Financial problems were greatly aggravated by workers being forced to spend wages in certain shops belonging to the mine owners and their families or friends.

Eventually, when Peñarroya was liquidated, a large part of the mining territory landed up in the hands of Portmán Golf. For a short while it existed as a mining company before going over wholly to construction. I have met men who worked for it in that period circa 1990. A lot of testing of ores seemed to be going on both from the Sierra Minera and distant places such as Lorca and Cabo de Gata.

There have been brief attempts to revive mining with modern methods, but Portmán Golf never seems willing to sell or lease the land for this purpose. During the property boom there was certainly more money to be made from using small sections of the land for other purposes. It's complicated though, as some is national parkland now and other parts need decontamination before they would be suitable for building. Mining Tourism could have been a possibility. While the renovation of Mina Agrupa Vicenta represents a laudable attempt to present mining history I would prefer another kind of mining tourism with some preservation of the buildings, lots of signposting and areas where collection of stones is allowed.

26: *The end of mining Lavadero Roberto plus Portmán beach*

Fiona Pitt-kethley

23: The Close Down of Mining

While many of the mines mentioned in this book had a short lifespan, the final close down of mining happened in 1991 to 1992. It has never reopened and perhaps it never will. The Sierra Minera was never a coal area and the world can do without coal by finding its heat and power in other ways. But the world cannot do without metals of which the Sierra Minera still has a rich supply.

Mining will continue to exist while the world does. It may be done in open quarries which are generally healthier for the workers though perhaps less healthy for the land. The film Blood Diamond made some people boycott diamonds because of the cruelty of the mining lifestyle. But these people cannot also boycott the rare metals in their mobiles or the copper wire in their electrical systems. There is equal cruelty in the mining of substances less luxurious than diamonds. Mining in many places is a cruel lifestyle but it does not have to be like this. There are ways of making it less hard and minimising contamination although this could also lessen profits.

Greed was at the heart of the death of mining in this area. Production levels were constantly stepped up and those involved ceased to care about contamination of the land. The technology used in the processing of lead and zinc, mainly, was impressive. Its ruined remains lie scattered amongst the countryside I love. If only techniques for decontaminating had kept pace this industry might still be alive.

At the height of production before the closedown there were only a few hundred miners involved. In Roman days there were forty thousand. In the early Twentieth Century, fifteen thousand were on strike.

In 1991, when Llano del Beal was saved from destruction by its militant inhabitants, their first industrial action involved shutting themselves in the Palacio Consistorial (town hall) in Cartagena. Eleven were injured in a confrontation with the police on the 12th of February. Over the months that followed various discussions between politicians and Portmán Golf went on. The final result was that Portmán Golf decided to close its mining operations and not reopen them until the price of metals went up by 50 per cent. This may have happened by now, but everything remains closed. Some land was requalified for building purposes. In reality the company is far more interested in that.

On the 22nd of October, MetalEurop, the proprietor of Peñarroya's factory in Cartagena announced a closedown. Peñarroya had been the country's first manufacturer of batteries and was producing 60 per cent of the country's lead. It had 352 workers.

On the 29th October, the miners kidnapped the commission's negotiator for some hours and days later made various verbal protests at the Asamblea. On the 4th of November, a group of miners set fire to chalets belonging to the two owners of Portmán Golf, Mariano Roca and Alfonso García. There was damage to the properties, but nobody was injured. The fires were put out quickly.

Alfonso García's main residence in the Plaza de España was already guarded so could not be attacked. A general strike was planned for December which was also in relation to the grave situation in other industries where thousands of jobs had been lost.

On the 14th of November, Portmán Golf announced its plans for compensating miners and developing the area. Just a few workers were kept on for a while in the offices and as security guards. Protests began the next morning with machinery brought out to block the road between La Unión and the Mar Menor, including a vast digger, six metres high, known as La Tonta and barricades of stones and earth.

On the 25th of November, the miners kidnapped politicians for seven hours, trapping them in the Asamblea. Three doors were blocked at twelve and nobody was allowed to leave. The trapped politicians scavenged what food they could in the building, unsure when the siege would end.

Over the months that followed more jobs were lost in industries which had some connection with mining, Zincsa, which now needed zinc from the Asturias rather than the local mines and several firms which produced fertilisers.

Sixty thousand workers were on strike on the 17th of December, a time when Cartagena's future looked bleak. With the huge decline in industry, other sectors were affected. If you are not earning you are unlikely to spend much on eating out.

On the 24th of January in 1992, Peñarroya in Santa Lucia, finally closed its doors with the loss of 350 jobs. Cartagena entered a period of great unrest. On the 3rd of February, during a big clash between protestors (mainly from the shipyard, Bazan, and Peñarroya) and police, forty people were injured as well as damage to property and cars nearby. Students from the University building opposite cheered protestors on. Smoke bombs and rubber bullets only caused various regroupings of the protestors. Some politicians were amongst the workers and other crowds from the closing fertiliser factories also joined the throng. Nearby schools had to be evacuated before tear gas was used. The Asamblea was set on fire by a Molotov cocktail thrown from the crowd. It is thought this was done by a group of young vandals rather than legitimate protestors. The fire was put out, but many photos still exist of black smoke pouring from the building. The area remained very sensitive and for many years special permissions were needed two weeks before for anyone displaying banners or protesting anywhere near the Asamblea.

In March 1992, Peñarroya workers made a chained protest at the Asamblea. PP proposed the expropriation of Peñarroya to keep their jobs. Other protests involved disrupting the railways and entries to the city. In May, they marched from Cartagena to Murcia.

Over the months that followed requalification of the land was discussed so that 17,000 houses and golf courses could be built. Some insistence on decontamination could have been made but wasn't. Some of the taxes payable when buying property could also have been devoted to this but the bay of Portmán remains as it is for the time being.

Everyone expected someone else to pay. From the point of view of Portmán Golf it should be the region or Madrid. The kindest interpretation that can be put on this is that the local authorities did not want to lose the construction work involved that could, at any rate, put some of the sacked employees of the area back to work. The only limits imposed were to respect the various tracts of beautiful countryside alongside, the Playa de Calblanque, the Monte de las Cenizas and Peña del Águila.

In December, Greenpeace asked for prison sentences for the four directors of Peñarroya. According to Greenpeace, the bay was contaminated up until 1990, with 7000 tons of toxic mud containing, lead, zinc, cadmium and silver. This killed off the marine flora and much of the fauna in the area. The legal processes were five years in the courts passing through the hands of four different judges. But nobody ended up in jail. The company had performed within the rather inadequate limits it had been set.

In April 1994, a fire threatened the older La Manga golf course. A thousand tourists and residents had to be evacuated from Peña del Águila. A few months later that year, it was time for Portmán Golf to start its huge building project.

27: *Calcination Oven, El Gorguel*

24: BIC

It was natural that many of the old mining areas should eventually become BIC, Bién de Interes Cultural. This gives them a preserved status without the money to repair them, or even to police them. It is also very much open to interpretation. The old mining buildings are not to be destroyed, that's obvious, but what of the hillsides? Is every stone and blade of grass to become untouchable if it all gets a BIC designaton? Collecting was a wonderful hobby and education for kids, and a source of tourism in the mining areas. Should that all be discarded? The rules need to be made plainer at the very least. One member of our association got a permit allowing him to collect samples for the local museum. I wanted to do something similar and asked at the town hall in La Unión. After some ringing round I was told as I belonged to this association, I could collect what I wanted at my own risk. I now carry a list of minerals the museum wants. In theory, I could supply them. I did indeed give some specimens to the smaller Las Matildes museum when it was open. Some others from my association are troubled and depressed by the thought of a BIC ruling and the interpretation given to it online by one Murcian academic who would like all mineral collecting to become illegal. One of my friends took down his own beautiful and informative website as a protest. Others fear the diminishing possibilities for collectors. Those who are more anarchic work round it. One told how the police had stopped him with his car boot full of rocks and a legón alongside. He refused to sign any statement of wrongdoing and said the stuff had been in his boot for years. There was nothing the police could do.

My love of history welcomes the idea of a BIC ruling for the preservation of the buildings. I can also see a case for some limit on how things are mined. Some tourist miners have come with power drills and left a terrible mess and a dangerous situation in the mines they have gone to. Local rumours tend to blame German mineral clubs for this because of their thoroughness. Hand tools seem preferable in such areas, giving more possibility to pick away at rocks with fine control and without leaving a bad situation behind.

Care needs to be taken in mines so that they aren't left in a dangerous state for others who follow after. On the hillsides there should be a freer situation. Taking a few dozen loose stones exposes other layers and does no more harm than picking a few figs off the nearby trees or mushrooms from the woods. Digging, however, needs to be done with care, again for those who come next. There should be no holes for them to fall into. Care also needs to be taken that rare plants are not removed.

Within the above limits, surely a compromise is possible and to everyone's advantage? Perhaps a yearly licence would also be a possibility with some insurance cover. Why should mineral collectors be treated worse than hunters? On the whole they are less of a danger.

The authorities were very selective at first as to which parts they labelled BIC. It was interesting to draw conclusions from what had been left out. On the surface there seemed no logic until you realised that areas that were not labelled BIC were perhaps ripe for development. Once you make somewhere BIC there is no going back. It is protected, and any number of local societies and individuals can bring denuncias against those who try to alter those buildings or that land.

Now that many mining remains in El Gorguel are BIC, in theory this should make the macro port impossible. But will it? Will some judge allow them to be over-ridden or use some legal quibble? When Monte Sacro was built on in the city a judge ruled that there was no proof that there was a city wall under the development site. Part was visible at two points, but the law did not allow that those two points were necessarily connected. I am sure similar judges could find similar loopholes to allow the destruction some politicians would like. There is always hope that the EU will over-rule them, which has happened in the past. The more evidence of archaeological remains, rare birds and plants that we the public can provide, the more chance of saving this beautiful area from destruction.

28: Polluted Mar Menor

Fiona Pitt-kethley

25: The Dying Mar Menor

It is not only industrial architecture that needs protection. When the wildlife in a renowned area of natural beauty dies it is more than a local matter. Looking at the reasons behind it is important on a national and international level. Lessons are there to be learned whether that area can be improved or not. It is a microcosm of what could happen to larger seas if we do not mend our ways.

The Mar Menor was a slice of paradise valued by inhabitants and visitors alike since Roman times. It has a higher level of salinity than the Mediterranean from which it is separated by a narrow strip of land known as La Manga. It was full of many different species including eels and sea horses. The fish and shellfish were prized by local restaurants.

The Mar Menor is now chocolate brown, and in some areas, stinks of rotten eggs. A huge storm known as the DANA was the tipping point, the straw that broke the camel's back. But it was definitely not the cause of the area's degradation. The blame lies elsewhere, largely with the politicians.

40 years or so ago, journalists began to hint at problems and a decline in water quality. Yet politicians failed to act to save it again and again. After the great

storm in September 2019, reports came in of small fish including sea horses washing up dead on beaches close to the end of the sea that is part of Cartagena, most particularly in Mar de Cristal and Playa Honda. This end was heavily developed during the building boom of the 90s and the early years of the 21st century before La Crisis slowed such projects. That was horrific enough but what was to come was worse. At the North end of the lagoon, away from the Sierra Minera, dead fish in even larger quantities appeared and, worst of all, they were seen leaping out of the water, gasping for air, only to die along the shore. The area of San Pedro and Lo Pagan was famous for mud baths, Dead Sea style. Nobody would risk the mud or the water now.

Groups that had been quietly urging politicians to do something became more militant. A small meeting was called in the district of Santa Lucia in Cartagena. It was literally at the bottom of the road where I live. Santa Lucia is a fishing district. It may also be the point where St. James landed in Spain. The fishers were ready to stand by the fishermen of the Mar Menor who are now unable to continue as there are few healthy fish left to catch. 150 families in San Pedro are now out of work. The effects on tourism are also not good. Those who had planned holidays there are thinking of going to other places where they can be sure of a good swim without coming out in a rash from entering toxic water. Many properties are empty or marked for sale without much hope of selling.

The meeting in Santa Lucia spawned a large protest on the 30th October 2019. A whole wide road in Cartagena was packed solid and we all processed slowly to the Regional Assembly building, that focal point once set on fire by the angry populace. Photographers were viewing us from cranes and the newspapers estimated the crowd as 55,000. As Cartagena has a population of round about 210,000 that is impressive. Looking at my fellow protesters there were certainly some that had never before attended marches or demonstrations. There were babies in pushchairs and octogenarians who could hardly walk making the effort for something they cared about. A few hundred protestors from Cartagena bussed to the large climate summit in Madrid in December. Strangely, Spain had failed to present anything about the Mar Menor and its problems. While we reminded some journalists of its existence we did not get as far as the point where Greta Thunberg was speaking, as half a million people were jamming the streets.

The Mar Menor is the subject of great affection. People rented summer houses along its shores or took the narrow gauge FEVE train down there for the day. The FEVE ends at Los Nietos which took a heavy hit in the storm. Los Nietos means grandchildren in Spanish. Many people here have memories of playing with children and grandchildren there in summer.

What makes the problem of the Mar Menor hard to fix is that it has several causes. A lagoon with high salinity has to keep its balance. One factor in losing this was the cutting of a canal in the La Manga strip so boats could enter. The seas were connected. It never was quite the same. The extreme foulness of the water now and its lack of oxygen has many other causes though. Modern farming is the biggest culprit. Huge tracts of land have been farmed aggressively, blasted with pesticides and then fertilisers added to every crop. These have washed down the old ramblas from as far away as Torre Pacheco. Years ago, the nitrates in the water were blamed for a huge rise in the jellyfish population and the Mar Menor began to be less of a paradise. Holidaymakers and residents also referred to the lower water quality as the "sopa verde" (green soup). Ecologists recognised it as a fall in oxygen in certain areas. Irrigation is one of the biggest factors. Any fresh water lowers the salinity. If the water is also contaminated, then that creates more problems. A hydrogeologist has raised an interesting theory that shows underground waters being wasted and running into the lagoon. Murcia has been perceived as dry and needing a lot of irrigation. Water is bought dearly from the companies that control it. Some is river water, the rest is desalinated. Many corruption cases have centred on the connections between politicians, water companies and desalination plants.

The ramblas running down to the water have proved disastrous.

A new route may need to be cut running to the Mediterranean rather than the Mar Menor, though this of course would not be great for the Mediterranean. Better still to find inland solutions to the problems. The battle for the sea needs to be won on land. I began to walk in the ramblas to see evidence from ancient mining. Sometimes healthy soil sits in a layer on top off something more contaminated. Sometimes purple granular NPK fertiliser appears on the paths alongside that edge the huge fields.

A quiet country road runs close to the coast and passes through farming land. A little further inland a motorway separates most of the mined part of the Sierra Minera from the coastal plains. There is no doubt that there are contamination problems from old tailings ponds. How much reaches the sea from the visible mines is a debatable point. A tiny percentage may indeed come from there, but I did not see much to indicate huge amounts on my walks in the ramblas and under the motorway. I think that the answer is simpler. The rambla that connects with Llano del Beal and El Beal is highly polluted as it leaves that area but becomes less so by the time it passes under the motorway.

After the September storm the tunnel area was relatively clean but the continuing rambla has been cut into heavily contaminated land. On either side you could see the old layers that turn up in tailings ponds and the trickle of water in the centre was polluted. The fields either side are farmed where a soil layer was added. The motorway acts as a kind of natural block to a great part of the mining pollution, but what lies on the side nearer the sea is more contaminated than most people realise. It has been in that state for a very long time.

My planting in nearby Lo Poyo with one of the big ecological organisations known as ANSE had shown me that area was already heavily contaminated. Most of what we planted were glassworts. They can grow in salty contaminated soil of this kind as can tamarisk. Lo Poyo was offered up for development with the land that includes the Monasterio de San Ginés. Hansa Urbana eventually did some rather basic repairs on the monastery but building plans got halted. It is currently suing Cartagena for many millions in compensation. I can only marvel at why the development was ever thought of as a good idea. The area needs to be returned to a humedal as should be many areas along the Mar Menor. Building and farming in areas which are already heavily contaminated with heavy metals from old mining processes should never be allowed again in the future. The authorities are beginning to realise something must be done but so far, their measures have been rather half-hearted. It is not enough to try to regulate some of the ramblas, wells and watercourses and ban pestcides within half a kilometre of the Mar Menor. Given storm conditions heavy chemicals can travel much further than that.

Other smaller local matters of ill-maintained storm drains and illegal irrigation systems also added to the disaster, but the problem is bigger. Another element is the provenance of sand for these beaches. Most of this has been swept into the sea and combined in the deadly cocktail. There were no natural sand beaches, but these were added alongside the developments along the coast. Some of the Mediterranean pebble beaches are also topped up yearly by the council. I have always disliked this practice and have a suspicion this sand might come from somewhere contaminated. The particles glitter like metal and I am reminded of the soil in quarries with chalcopyrite and pyrite, something that could add acidity to the water. It is another point that needs to be checked out by the scientists.

55,000 people got out and showed they cared about the much-loved Mar Menor at the end of October. There has already been some limiting the type of farming in the area between the sea and the motorway. Even ecological farming and crops that do not require irrigation may be too much for the Mar Menor. I would suggest that a whole section should be turned over to nature and the replanters. There are many dozens of us willing to participate in such schemes and I am one of them. There could be cycle paths and fixed routes through this flat wasteland. It would certainly appeal to bird-watchers and other eco-tourists. With this sort of barrier in place the sea might have time to recover. ANSE has already bought a small part of the land and water of the Salinas de Marchamalo which used to be part of the salt trade and that is something hopeful. This is close to the entrance of the La Manga strip.

Should the canal be filled in? It is already closing itself with silt. But La Manga has become used to the passing of boats though this narrow gap. La Manga has lived on tourism but is dead in winter. Tourism may have helped kill the golden goose while aggressive farming finished the job.

There are a few birds picking at the edge of the chocolate-coloured water in Los Nietos these days. Gulls are known for eating any old rubbish, but I also saw an egret. The birds were finding some life even if it was only dying mosquitoes. I compared them in my mind to old men visiting a bar they had loved that was about to be closed down for hygiene reasons. They were living on the memories of what it once was.

29: Calcination Oven near El Lirio

Fiona Pitt-kethley

Glossary: Minerals of the Sierra Minera

The Sierra Minera is known amongst mineralogists for its vast variety of minerals. I have divided the list into two parts. The second half concentrates on micro minerals. These usually need to be identified with a microscope. Amongst this list I have also included some very rare minerals and those that only exist in small quantities. Between the two lists I have included everything that has been found on the Sierra Minera. In some cases specific minerals have not been found for several decades, No one has managed to collect all these, not even the mineral museum in La Unión. It is probable also that other minerals will be discovered in the area with improvements in analysis. Deviline is a recent find. There are probably also discoveries to be made if anyone cares to analyse the sand in the Mar Menor. The Romans are said to have dumped a lot of mining waste there.

I have included a few basic details with each mineral: the chemical formula, hardness on the Mohs scale and the type of crystal formation, sometimes its colour also.

Minerals

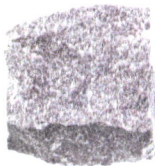

Alunite $KAl_3(SO_4)_2(OH)_6$ Hydrated aluminum potassium sulphate. Hardness 3.5 to 4. Trigonal. It occurs chiefly in Los Blancos and Cabezo Rajao. It was mined from ancient times. It is one of several slightly different minerals known as alum. It was also mined in Alumbres, giving the village that name, alumbre being the Spanish for alum.

Amethyst SiO_2 Silicon dioxide. Hardness 7. Its form is hexagonal, sometimes biterminate, sometimes in clusters of crystals. Its purple colouring is usually caused by impurities from iron or manganese. Sierra Minera amethysts are mostly pale shading to darker at the tip. The easiest place to find them is between La Unión and Llano del Beal in the slag tips and on hillsides and more occasionally in Portmán. Larger amethyst geodes have been found hidden in the large brown boulders of the San Valentín quarry. Pale amethyst is also found mixed with zinc blende and galena in Túnel José Maestre. Unlike most of the minerals in this list, which were solely a side product from the search for metals, amethyst was mined for its own sake. A concession for mining them and sapphires existed at the end of the sixteenth century. Probably these were not sapphires by current usage of the word. If so, none have been found since. Mineral identification has made great strides in the last two centuries. Before that many types of stones were lumped together. This concession was in the Portmán area.

Anglesite PbSO$_4$ Lead Sulphate. Hardness 2.5 to 3. Specimens in the Sierra Minera are mostly white, colourless, grey or yellow. Its form is orthorhombic. They most often occur in large yellowish blocks of stone in the quarries alongside other minerals like cerussite and galena. It occurs alongside galena in Mina Descuido with mimetite and cerussite. It also appears in the Cuesta de las Lajas in Mina Relámpago and the quarries of San Valentín, Tomasa, Emilia and Los Blancos. It is also in El Algar. Most specimens these days are small and not high quality. There are larger specimens from decades ago in mining museums.

Antimonite see **Stibnite** in Micro Minerals for more info

Aragonite CaCO$_3$ Calcium carbonate. Hardness 3.5 to 4. Most commonly cream coloured except when it is stained orange by iron. It's present in many cave type mines in the Sierra Minera. The most notable formations are in Cueva Victoria and the side passage of the María Dolores mine These contain whole walls covered with rippling formations of aragonite. The mine Los Aragonitos also contains some broken chunks of it right at the end of its passage. It is found in coraloid form in mines of the Collado de Tinaja and in Mina Pepito and Mina Rómulo. It is on calcite stalactites in the Santa Isabel mine on Collado de Don Juan.

It is also in massive form in the Tomasa quarry and is also present in Mina Blanca and Mina San Antonio. Long crystals have been found in the Cabezo de San Ginés area. These have a pyramid shape and are yellow under ultra violet light. Micro crystals are present with other minerals in Mina Marisol.

Arsenopyrite FeAsS Iron arsenic sulphide. Hardness 5.5 to 6.4. Monoclinic. Greyish-black. It occurs in Mina Rosa and in greenalite masses and in the marcasite of the San Valentín and Gloria quarries. It is also alongside pyrite and marcasite and pyrrhotite in the Brunita quarry.

Azurite $Cu_3(CO_3)_2(OH)_2$ Copper carbonate. Deep blue. Hardness 3.5 to 4. Monoclinic. Mostly occurs as very small crystals, often alongside malachite. There is a lot of it in the San Valentín quarry, often alongside galena and without malachite. It has also been found in the mines known as Inglesa, San Juan Bautista, Belleza, Paraíso, Paulino and Agradecida. The largest crystals are only found by the mine known as La Verdad de Un Artista.

Barite $BaSO_4$ Barium Sulphate. Hardness 3 to 3-5. Orthorhombic. The barite of Cartagena is famous in mineral circles and comes in many different forms. Colours vary from white, black and white, orange, blue and green. Barite is one of the most abundant minerals in the Sierra de Cartagena. The only barite mine was in El Abrevadero but it is everywhere in Cerro de San Ginés, Ramblizo de las Nogueras, Cabezo de Ponce, El Gorguel and the Campos de Golf. Pure white crystals occur in Alumbres and Vista Alegre in the San Camilo mine. Some include purple fluorite. Large black and white specimens occur in El Estrecho and the mines of the Campos de Golf.

Mina Haití has delicate dark specimens set in limonite in the ceiling of the lower levels and some greenish specimens as well, Mina Teresita, Mina Marisol, Mina San Timoteo and Mina San Jorge have blue barite. The San Jorge mine is also known as La Carpeana. There are very delicate specimens of white barite like snowflakes in the Santa Bárbara mine in El Gorguel and in the side passage, about half a kilometre along Túnel José Maestre.

Fiona Pitt-kethley

La Gaviota in Cola de Caballo has barite specimens with rare micro crystals of other minerals alongside. It is found with siderite in the Brunita quarry and with greenalite in the San Valentín and Tomasa quarries.

Bornite Cu_5FeS_4 Copper Iron Sulphide. Hardness 3 to 3.5. Black tarnishing to iridescent. Orthorhombic. It occurs in Corta Brunita, sometimes alongside pyrite and marcasite.

Calcite $CaCO_3$ Calcium carbonate. Hardness 3, Hexagonal. It has the same chemical composition as Aragonite but differs in structure. Mostly white in the Sierra Minera unless stained. It occurs separately and also in thin layers over quartz and as stalactites. It's widespread between Llano del Beal and La Unión and on Cabezo de San Ginés.

There are stalactites in the Túnel José Maestre in Portmán where it combines with zinc blende. The best examples of calcite are those found in the Campos de Golf in the mines Manolita and Herculano. Some good specimens have also been found in Mina Esperanza and Mina Haití and other mines of the Ramblizo de las Nogueras. It occurs on the lower level of Mina Precaución also and in Mina Victoria. In many mines between Llano del Beal and La Unión a secondary crystallisation of calcite on quartz is rather less beautiful and people remove it with acid.

Cassiterite SnO_2 Tin oxide. Hardness 6 to 7. Tetragonal. It occurs in the Cuesta de Las Lajas and Paraje de Quebrarados in La Unión where tin was mined. Greyish, brownish. Few specimens these days. It was mostly found close to Mina Remunerada. The mines San Isidoro and Marinera have produced specimens. These have also been found in the spoils from Mina Cuarta and in the rambla Los Cucones by Alumbres and the Paraje de Las Pocilgas close to La Esperanza.

Celestine $SrSO_4$ Strontium sulphate. Hardness 3 to 3.5. Orthorhombic. Blue. It is uncommon but has been found in one section of the Tomasa quarry. The blue crystals have micro crystals of calcite alongside.

Cerussite $PbCO_3$ Lead sulphate. Hardness 3 to 3.5. Orthorhombic. White. It is widespread in the Sierra Minera though mostly small poor specimens with other minerals. It occurs in the Las Lajas and Sancti Spíritu areas, in the Vulcano, El Humo, La Superior, Inglesa, Descuido, San Antonio, Los Negros, El Corcho, Belleza, San Juan Bautista, Fortuna and Tercera España mines, it is also in mines of La Crisoleja and Cabezo de Ponce. It is in the San Valentín, Emilia, Los Blancos, Sultana and Brunita quarries. It often occurs alongside anglesite.

Chalcedony SiO_2 Silicon dioxide is a form of quartz. Hardness 7. Hexagonal. Most often white or blue, but occasionally other colours. It often occurs with quartz or opal in the Sancti Spíritu area and in the San Valentín, Tomasa quarries. It is occasionally found in the Gloria quarry also.

Chalcopyrite $CuFeS_2$ Copper iron sulphide Hardness 3.5 to 4. Tetragonal. Golden sometimes tarnishing to rainbow colours. Chalcopyrite chiefly occurs in the Brunita quarry, with smaller micro specimens in San José and Gloria and Mina Gaviota on Cola de Caballo. The sands of that beach have a large proportion of chalcopyrite, making them unbearably hot in summer.

It also occurs in Mina Trinidad and the mines of Cabezo de Don Juan and Sancti Spíritu.

Cinnabar HgS Mercury sulphide. Hardness 2 to 2.5. Red. Hexagonal. It is present in the Sierra de la Fausilla in Escombreras. It is visible alongside the Emasa quarry used for building materials.

Citrine SiO_2 Silicon dioxide. Hardness 7. Hexagonal. Yellow. It is also known as burnt amethyst. Small quantities of it occur in the hills between Llano del Beal and La Unión. There are also yellow-stained pieces of quartz which can be confused with it. The true citrine in this area is in clusters of crystals on a metallic matrix that looks greenish from behind.

Copper Cu. Hardness 2.5 to 3. Isometric. Native copper has been found in the south west of the San Valentín quarry in geodes and fissures. Some is on show in the La Unión museum. It has also been found in the Belleza, Paulino and Agradecida mines of Sancti Spíritu.

Cuprite Cu_2O Copper oxide. Cuprite has been found in small quantities alongside malachite, azurite, linarite and native copper in the San Valentín quarry in the area where the San Juan Bautista mine was. It has not been found for a long time.

Dolomite $CaMg(CO_3)_2$ Calcium magnesium carbonate. Hardness 3.5 to 4. Hexagonal. Mostly white. It occurs with Ankerite and Calcite close to Mina Julio César and Mina Buen Consejo. A violet crystalline form is found in geodes above the Brunita quarry. Under fluorescent light it is rose coloured. It is also found in Paraje del Huerto de San Pedro between La Unión and Alumbres.

Fluorite CaF_2 Calcium fluoride. Hardness 4. Isometric. It is mostly violet in the Sierra Minera. It loses colour with contact with sunlight. There is also some colourless fluorite. It is most often in cubic form. It occurs chiefly in Alumbres, sometimes alongside barite. The San Camilo mine, which is also known as Vista Alegre or Cartagenera, from the nearby foundry, has fluorite alongside barite and sometimes, bournonite. It has also been found in the Marisol mine in the Campos de Golf.

It is now far less common in Alumbres due to the expansion of the road to Escombreras, which caused mines in the area to be demolished or buried. It has also been found on the beach at Portmán and close to the port of Escombreras. With violet fluorite it is necessary to keep it away from light as the colour can fade.

Galena PbS Lead sulphide. Hardness 2.5 to 3. Isometric. Steel grey metallic. It is one of the commonest minerals in the Sierra Minera. It was exploited chiefly in the mines of Cabezo Rajao and of the Barrancos de Mendoza and de Francés and in the mines Laberinto and El Arresto in El Gorguel. It is also in Mina Lolita. It occurs in blocks of greenalite in the San Valentín quarry sometimes with smithsonite. I have seen a spectacular crystalline galena formation from the walls of San Valentín.

There is a beautiful geode of quartz covered with galena in a very small space in Mina Rómulo. It is found with quartz in the mines of Collado de Don Juan. There is also much galena with the zinc blende in Túnel José Maestre. At times it is easy to confuse them. It has also appeared in Corta Sultana in druses with zinc blende, in Corta Brunita with pyrite and in Los Blancos. Large geodes came from the latter quarry at one period.

It is also in Mina Segundo Ferrocarril in large flat-looking crystals. One of the main centres for processing it was the foundry in Santa Lucia, part of the now deserted Peñarroya complex. The last manager there had a bin of huge chunks of galena to give away to interested visitors.

Goethite FeO(OH) Iron hydroxide. Hardness 5 to 5.5. Orthorhombic. Black, sometimes rainbow-coloured. It can appear as a streak in opal. It's abundant on Ruta 33, La Crisoleja, Sancti Spíritu and in the quarries of San Valentín, Emilia, Tomasa and Los Blancos as well as the beach at El Gorguel. It is in Mina Haití on stalactites and with barite in manganese mines of the Ramblizo de Las Noguera. It's prettiest in rainbow form. Examples have been found on Cabezo de San Ginés and in the mines Mercurio and Enrique. Because most of the rest is black it is hard to spot by a casual collector, so is a poor seller at mineral fairs. There are some good micro rainbow specimens in the San José quarry. It is attractive in its rainbow form as a thin layer on top of quartz or other crystals.

Gold Au. Hardness 2.5 to 3. Isometric. Gold is rare but exists in minute quantities, more traceable by analysis than the naked eye, alongside silver and lead in Cabeza Rajao.

Greenalite $(Fe^{2+},Fe^{3+})_{2\text{-}3}Si_2O_5OH_4$. Hardness 2,5. Monoclinic. Dark grey green. It forms a large part of what is known as the manto de azules, the silicate layer that tops the Sierra Minera. It occurs on the Collado de San Juan and in the quarries of San Valentín, Emilia, Tomasa, Gloria, San José and in Llano del Beal. Often, other minerals are visible in large rocks formed of it, galena, sphalerite, smithsonite, siderite and quartz. Broken into small chunks it has been used along the FEVE railway line and powdered into the road surface between Llano del Beal and Portmán.

Greenockite CdS. Cadmium Sulphide. Hardness 3 to 3.5, Hexagonal. Found near the Los Blancos quarries in a yellowish form, it occasionally forms large crystals with sphalerite- Often, it forms sulphur-like patinas of colour in some mines near Portmán such as San Antonio, Lola and Túnel José Maestre. Only chemical analysis can differentiate this from ordinary sulphur. I believe this is partly responsible for the yellow colour of the flood water in Túnel José Maestre as the colour takes more than a day to wash off your skin.

Gypsum $CaSO_4 2H_2O$ Hydrous calcium sulphate. Hardness 2. Monoclinic. Most specimens are white or colourless except for the acicular ones which come in pink or orange. It is widespread in various forms in the Rambla del Moro in Portmán and Cabo de Palos. Gypsum was mined near Cala Cortina and I have found it in glasslike form in Mina San Águstin there. It is also in the Gloria, San Valentín and Tomasa quarries. The most attractive forms mostly come from Portmán and especially Mina San Timoteo, where they are acicular and very delicate. There is a delicate gypsum variant knowns as sericulite in Mina Gaviota. The largest gypsum crystals were found back in the 70s in the Iberia mine on the fourth level at 240 metres deep where it joins with Mina Artesiana. Large crystals were also found in the Sancti Spíritu area and in Mina Dificultad. Mina Bragelone, Mina Balsa Depositaria and Mina Segundo Ferrocarril have also yielded good specimens.

Halite NaCl Sodium chloride. Hardness 2 to 2.5. Isometric. White. Halite is found in the Salinas of Marchamalo and Calblanque.

Hematite Fe_2O_3 Iron oxide. Hardness 5 to 6. Hexagonal. This occurs in the form of specularite on Monte San Julián in the slag heaps above the entrance to Mina San Águstin by Cala Cortina. Specularite is dark grey and silvery and leaves metallic dust on hands. Hematite also occurs in the mines of Los Cucones and El Humo and on Cabezo de San Ginés with braunite, manganite, todorokite and birnessite.

Hemimorphite $Zn_4Si_2O_7(OH)_2H_2O$ Basic hydrous zinc silicate. Both hemimorphite and smithsonite were often lumped together by miners under the name calamine. Hardness 4.5 to 5. Orthorhombic. Usually white. Hemimorphite occurs chiefly in Mina Precaución on a matrix of limonite though there is some in Cabezo San

Ginés, the San Valentín quarry. Corta Sultana and mines and roadside at Portmán. The best specimens come from Mina Precaución.

Jarosite $KFe^{3+}_3(SO_4)_2(OH)_6$ Basic hydrous potassium iron sulphate. Hardness 2.5 to 3.5. Hexagonal. Jarosite has been found in Cuesta de Las Lajas and Los Blancos and also on limonite in the San José and San Valentín, quarries and in fissures in Cabeza Rajao.

Lead Pb. Hardness 1.5. Isometric. Grey metallic. Native lead only turns up very occasionally in clay.

Limonite $FeO(OH) \cdot nH_2O$ Hydrous basic iron oxide with an indefinite composition. Hardness 4 to 5.5. Amorphous. Orangey-brown and very porous. It's widespread in the Sierra Minera and is the matrix to some other minerals such as the hemimorphite in Mina Precaución. It's also widespread in the quarries and Cabeza Rajao.

Ludlamite $(Fe,Mn,Mg)_3(PO_4)_2 \cdot 4H_2O$. Hardness 3.5. Monoclinic. Green. It occurs in Corta Brunita, most often in geodes with pyrite, quartz or siderite. Occasionally it is alongside vivianite.

Magnesite $MgCO_3$ Magnesium carbonate. Hardness 3-5 to 4. Hexagonal. It is said to exist in the Sierra Minera but I can find no precise locations mentioned.

Magnetite $Fe^{2+}Fe^{3+}_2O_4$ Hardness 5.5. to 6.5. Isometric. Black. Magnetic. It is common alongside greenalite. It's abundant in the Tomasa quarry and has also been found in Corta Emilia alongside quartz and in Corta Brunita. It is abundant at a microscopic level in the limestone of the Cabezo de los Cuervos. It is generally associated with greenalite. In the mines of Sancti Spíritu it is found alongside pyrite.

Malachite $Cu_2CO_3(OH)_2$ is widespread in the Sierra Minera. It is often associated with azurite but this doesn't seem to be the case in the San José quarry where it is found alone in micros. It is found in Mina Amable in the Paraje del Cabezo del Pino, Portmán and Mina Consuelo. It is present in the quarries of San Valentín, San José and Gloria. Very poor-quality malachite turns up also behind the Peñarroya buildings in Santa Lucia.

Marcasite FeS_2 Iron sulphide. It has the same chemical formula as pyrite but crystallises differently. Hardness 6 to 6.5. Orthorhombic. Yellowish metallic tarnishing to brown. It occurs in the mines of Cabezo del Pino, Cabezo Rajao and the Rambla de Mendoza. It is also in tiny points alongside marble and barite in the Tomasa quarry. It has been found on quartz crystals in Corta Brunita and in the San Valentín quarry. It is also found with pyrite in the Los Blancos quarries and in the mines Buen Consejo and Julio César.

Marmatite see Sphalerite.

Melanterite $FeSO_4 \cdot 7H_2O$. Hydrous iron sulphate. Hardness 2. Monoclinic. Blue green. It occurs in pyrite mines in Cuesta de Las Lajas such as Remunerada and Pablo y Virginia. It is also in Cruz Chiquita, Aries and El Tesoro mines and in Calblanque. It occurs in the Brunita quarry too.

Mercury Hg. It has been found in Mina Arresto in El Gorguel.

Monheimite see Smithsonite.

Opal $SiO_2 \cdot nH_2O$. Hardness 5.5. The commonest colours are yellow ochre and a deep red. These two are often referred to by a separate term in Spanish "jaspe" which is not to be confused with jasper. They have a higher iron content than some of the other colours. They are very common in the Sancti Spíritu area. A path near the San José quarry contains a lot of blue opal alongside chalcedony. There is also cream, brown, purple and black opal. In the San Valentín quarry there is green opal. Opal also appears in the Emilia and Tomasa quarries and in the La Crisoleja and de los Quebrarados area of La Unión. Some of the finest specimens of opal are known as "porcelainitas" for their extra shiny quality. Some striped opals resemble tiger's eye.

Prasiolite SiO_2 is a green form of quartz. Silicon dioxide. Hardness 7. Hexagonal. Greyish green. It was common in a part of Corta Emilia until the quarry was filled with inert residues. There is now too much invigilation there for mineral hunters. It is still found on a hillside backing on to the quarry, near the Segunda Esmeralda mine. It is also in the slag heaps by Mina Catón and is found occasionally in the San Valentín quarry.

Pyrite FeS_2 An iron sulphide. Hardness 6 to 6.5. Isometric. Brass colour. It's widespread in the Sierra Minera. It does not form large cubes like those known from Navajún but tiny ones sometimes occur. It is more often seen as streaks in other rocks. It is present in the quarries and also in the Cuesta de Las Lajas area and its mines. Mina Agrupa Vicenta was a pyrite mine as were some of those nearby. It was mined extensively and larger octahedrons were found several decades ago in Cabezo Rajao in the Iberia and Artesiana mines. These turned up more occasionally in the mines of San Rafael, La Belleza, Corta Tomasa and Corta Brunita. And the Santa Teresa mine in El Abenque. It occurs alongside pyrrhotite in Corta Brunita and occasionally as small cubes there and in the Los Blancos and Sultana quarries. It was mined with explosives. In the days when children worked at the rock face they were never allowed to work in these mines because of this.

Pyrrhotite. $Fe_{(1-x)}S$ (x = 0 to 0.2). Pyrrhotite is also known as magnetic pyrite because it is similar to pyrite but weakly magnetic. Hardness 3.5 to 4.5, Monoclinic. Bronze coloured. It is an iron sulphide. It exists in several quarries alongside pyrite, galena and sometimes zinc blende. It is commonest in the Brunita quarry where it is sometimes associated with vivianite and ludlamite. It has also been found in Túnel José Maestre and the Sultana and Los Blancos quarries.

Pyrolusite MnO_2 Manganese dioxide. Hardness 6 to 6.5. Tetragonal. Dull metallic. It occurs on Sancti Spíritu and in Cabezo de San Ginés, particularly the Haití, Africana and Marquesita mines and in dozens of other mines in Llano del Beal, El Abenque, El Estrecho and the Barrancos del Francés and de las Pocilgas. It is also common in the mines of the Campos de Golf. It's the matrix for calcite in Mina Herculano. It occurs with barite in Mina Marisol and Mina Joaquína. It is also in Mina Teresita with blue barite.

Pyromorphite $Pb_5(PO_4)_3Cl$ Lead chlorophosphate. Hardness 3.5 to 4. Hexagonal. It occurred in various colours, grey, green, yellow and white alongside galena in the Los Blancos quarries. The matrix is often ochre or goethite. It is sometimes with cerussite. It has also been found in the San Valentín quarry. Pyromorphite also occurs in stalactite form in Mina Gloria.

Quartz SiO_2 Silicon dioxide. Hardness 7. Hexagonal. Mainly colourless unless stained with contaminants, occasionally milky. Amethyst and prasiolite have their own separate entries. It is commonest in the mines and hillsides between Llano del Beal and La Unión. It appears in sceptre like crystals near the lake in Corta Brunita and in micro form in the San José quarry. There are geodes in the San Valentín quarry.

Some of the best specimens have come from the María Dolores mine. Smoky quartz appears in this area and has also been found in Mina Mercurio. Some quartz with crystals within crystals has been found in the La Parreta mine. Small crystals forming druses have been found in the mines San Antonio and Lola and in Túnel José Maestre.

Rodochrosite $MnCO_3$ Manganese carbonate. Hardness 3.5 to 4. Hexagonal. Pink. It occurs in the Peña de Águila area close to the San Rafael mines and Secretaría and Vibora. It is in the slag heaps. It sometimes occurs with siderite. It is also in the mine Carolina La Doncella and the San Valentín quarry. It is also known in the mines of Cabezo de San Ginés such as Mina Victoria and Joaquína where it is associated with galena and sphalerite.

Romanechite $(Ba,H_2O)_2(Mn^{+4},Mn^{+3})_5O_{10}$ is the primary constituent of psilomelane. Hardness 6 to 6.5. Tetragonal. Brown. It is found in the mines of the Campos de Golf and also in Mina Esperanza on pyrolusite. Its form is usually botryoidal.

Scorodite $FeAsO_4 \cdot 2H_2O$ A common hydrated iron arsenate mineral. Hardness 3.5 to 4. Orthorhombic. Greenish. When heated it smells of garlic. It weathers to limonite. It is widespread but not much collected.

Siderite $FeCO_3$ Hardness 3.5 to 4. Hexagonal. It is found alongside silicates in the Sierra Minera. Colours vary from blackish to cream. There is a particularly attractive caramel-coloured variety in the Brunita quarry. There are interesting pieces mixed with barite and chalcopyrite in the side passage of Túnel José Maestre. Micro specimens are best seen under a microscope where their folded geometry emerges. It is also common in the mines of Cabo de Palos such as Candida and Ferruginosa and in geodes in the mine Carolina La Doncella. It is also in the quarries of San Valentín, Tomasa and Gloria. In Tomasa it is associated with crystals of anglesite.

Silver Ag. Hardness 2.5 to 3. Isometric. Native silver has been found in the mines of Sancti Spíritu, La Crisoleja, Mina Belleza, Serrano, El Corcho, La Carolina, La Murciana and others. I also met someone who had been given permission to drill deep on Cabezo Rajao and had found some there.

Smithsonite $ZnCO_3$. Zinc carbonate. Hardness 4 to 5. Hexagonal. Usually white or brown in the Sierra Minera. It occurs in Mina Tetuán in La Unión and in the La Parreta area of Alumbres. Green forms have been found in San Valentín alongside copper minerals and on chalcedony in the form of Monheimite, a smithsonite with a high iron content. It is found in a brownish colour alongside quartz in the Brunita quarry. It used to be known as calamine.

Other mines include Santa Teresa, Iberia, Dichosa, Inocente, Lola, Don Quijote, Sancho Panza, Julio César, San José, Perdida and Buen Consejo. It is also in the quarries of Emilia, San Valentín, Tomasa, Brunita and Los Blancos. It is sometimes seen coating quartz in the Brunita quarry.

Sphalerite or **Zinc Blende** ZnS. Zinc sulphide. Hardness 3.5 to 4. Isometric. It is common in the mines of Portmán and La Unión. The most beautiful crystalline specimens chiefly occur in Túnel José Maestre and also from Mina Lolita. At one-time good specimens of it on top of Galena were found in Corta Sultana, but none in recent years. Other mines where it occurred are El Arresto, Lola, Ocasión, San Rafael, Lo Veremos, Secretaría, La Loba, El Huerto, Artesiana, Belleza, Trinidad. Many specimens are small and easy to mistake for other minerals. It was mined to the point of exhaustion in mines such as San Isidoro and Iberia on Cabezo Rajao. It has also been found in the quarries of San Valentín, Los Blancos and Brunita. Where sphalerite is iron rich it is known as marmatite and has a darker colour.

Stannite Cu_2FeSnS_4. is a sulphide of copper, iron and tin. Hardness 4. Tetragonal. Stannite occurs alongside greenalite and is included with sphalerite in specimens from the mine Julio César and others in the Rambla de Mendoza on the outskirts of Llano del Beal.

Sulphur S. Hardness 1.5 to 2.5, Orthorhombic. Some sulphur is visible in powdery form close to iron mines in La Unión and is perhaps on the beach at El Gorguel. Although this could be Greenockite which looks similar. Sadly it never seems to take the beautiful crystalline forms which appear in some mines in Lorca.

Vivianite $Fe_3(PO_4)_2 \cdot 8(H_2O)$. Hydrous iron phosphate. Hardness 1.5 to 2. Monoclinic. It was found in Corta Brunita associated with siderite, pyrite, pyrrhotite and ludlamite. The blue-green vivianite of Brunita is famous and some of the best in Spain. Unfortunately none has been found for a long time. Specimens need to be kept out of the light as light causes the colour to darken in time to black. There have been examples found in Los Blancos also.

30: El Laberinto

Micro Minerals

Micro Minerals and others that are extremely rare or have only been found in small quantities

Acanthite Ag_2S. Hardness 2 to 2.5. Monoclinic. Black. Found in micro form amongst waste in the San Valentín quarry alongside galena, chlorargyrite, stephanite and dyscrasite.

Adamite $Zn_2(AsO_4)(OH)$ Basic zinc arsenate. Hardness 3.5. Also known as adamine. Small orthorhombic crystals of a greenish colour have been found in geodes from the limestone of Mina Consuelo, Cartagena and in Mina La Verdad de Un Artista.

Aluminacopiapite $AlFe_4^{3+}(SO_4)_6(OH)_2.20H_2O$. Triclinic. It occurs in the Los Blancos quarry alongside Pickeringite and Halotrichite.

Amblygonite $(Li,Na)AlPO_4(F,OH)$ a fluorophosphate mineral. Hardness 5.5 to 6. Triclinic. White. It occurs in Cabezo de Don Juan and the mine Santa Lucia on the extreme north of that hill.

Anatase TiO_2 Titanium oxide. Hardness 5.5 to 6. Tetragonal. It occurs in Cuesta de Las Lajas, the Los Blancos quarries and in the beaches of Calblanque. Its colour is dark blue and it can be seen at a micro level alongside quartz, brookite and barite from the Los Blancos quarry. It has also been found close to the José Maestre tunnel and between that area and Mina Difficutad and in the El Sabinar quarry close to the Finca Primavera in Los Belones.

Andalusite Al_2SiO_5 An aluminium neosilicate mineral. Hardness 6.5 to 7.5. Orthorhombic. Black. It occurs in andesite close to the coast. It has been seen at El Carmolí.

Ankerite $Ca(Fe,Mg,Mn)(CO_3)_2$ Hardness 3.5 to 4. Trigonal. Brown. This has been found in the Brunita quarry with galena and zinc blende. It was also in the Julio César and Buen Consejo mines with barite. In the San Antonio mine it was found with greenalite.

Antimony Sb. Hardness 3 to 4. Hexagonal. Greyish white metallic. There are tiny amounts of antimony in the Los Blancos quarries and alongside galena in Mina San Rafael.

Argentojarosite $AgFe^{3+}_3(SO_4)_2(OH)_6$ Hardness 3.5 to 4.5. Trigonal. It is found at micro level in El Estrecho de San Ginés.

Arrojadite $KNa_4Ca(Mn,Fe)_{14}Al(PO_4)_{12}(OH)_2$ Hardness 5. Monoclinic. It occurs with Dickinsonite in the Los Blancos quarries in masses of jarosite and natroalunite.

Atacamite $Cu_2Cl(OH)_3$ Hardness 3 to 3.5. Orthorhombic It occurs on barite from La Gaviota together with embolite. It is green and crystalline.

Aurichalcite $(ZnCu)_5(CO_3)_2(OH)_6$ A carbonate of zinc and copper. The ratio of zinc to copper is about 5:2. Its form is orthorhombic. Hardness 2. Colour blue-green. It is rare in the Sierra Minera but occurs in small quantities in the San Valentín quarry with other copper minerals.

Bertrandite $Be_4Si_2O_7(OH)_2$ Beryllium Sorosilicate hydroxide. Hardness 6 to 7. Orthorhombic. It appears as a micro alongside greenalite in the Gloria quarry.

Beudantite $PbFe_3(OH)_6SO_4AsO_4$ Hardness 3.5 to 4.5. Trigonal. Brown. It exists in micro form in the Cuesta de Las Lajas alongside anglesite. And in the Los Blancos quarry

Birnessite $(Na_{0.3}Ca_{0.1}K_{0.1})(Mn^{4+},Mn^{3+})_2O_4 \cdot 1.5H_2O$ Hardness 1.5. Monoclinic. Black. Found at micro level on Cabezo de San Ginés

Boulangerite $Pb_5Sb_4S_{11}$ Lead antimony sulphide. Hardness 2.5. Monoclinic. Grey It is found in micro form with galena and greenalite.

Bournonite $PbCuSbS_2$ A sulphantimonite of lead and copper. Hardness 2.5 to 3. Orthorhombic. Grey. It exists in micro form alongside galena in the Sierra. It has been found with barite and fluorite in Mina San Camilo in Vista Alegre.

Botryogen $MgFe^{3+}(SO_4)_2(OH) \cdot 7H_2O$ Hardness 2 to 2.5. Monoclinic, Orange. It is found in the mines Tesoro, Belleza and Iberia.

Braunite $Mn^{2+}Mn^{3+}_6[O_8|SiO_4]$ Hardness 6 to 6.5. Tetragonal. Black. It is found at micro level on Cabezo de San Ginés.

Brookite TiO_2 Titanium Oxide. Hardness 5.5 to 6. Orthorhombic. Appears in tiny honey-coloured crystals alongside rutile and anatase in barite from the Los Blancos quarry and with anatase in various sites near La Unión.

Caledonite $Pb_5Cu_2(CO_3)(SO_4)_3(OH)_6$ Hardness 2.5 to 3. Orthorhombic. Blue-green. It appears alongside Linarite. It is in a museum piece in the United Kingdom but the exact location it came from is not given.

Cervantite $Sb^{3+}Sb^{5+}O_4$ Hardness 4 to 5. Orthorhombic. Yellowish. It occurs at micro level alongside oxides of iron in the mines of the Loma del Teniente.

Chalcanthite $CuSO_4 \cdot 5H_2O$ Hardness 2.5, Triclinic. Water soluble, so easily destroyed. Blue-green. It has been found in the Brunita quarry and in the mines Tesoro, Belleza, Cruz Chiquita and Iberia.

Chalcocite Cu_2S Copper Sulphide. Hardness 2.5 to 3. Orthorhombic. Grey. It exists in micro form in the Brunita quarry alongside quartz.

Chalcophanite $(Zn,Fe\,Mn)Mn_3O_7 \cdot 3H_2O$ Hardness 2.5, Trigonal. It is found in the manganese mines of Cabezo de San Ginés and in Mina Pepito and others in the Ramblizo De Las Nogueras. It gives a bluish look to the minerals. It is also in Mina Precaución. It occurs also in Mina Herculano with pyrolusite and in the mines Teresita, San Timoteo, Marquesita Moderna and Tercera Esperanza.

Chamosite $(Fe^{2+},Mg)_5Al(AlSi_3O_{10})$ A hydrous aluminium silicate of iron. Hardness 3. Monoclinic. It has been found in Corta Sultana.

Chlorargyrite $AgCl$. Silver chloride. Hardness 1 to 2, Isometric. Is found in the La Crisoleja area of La Unión and in the Los Blancos and San Valentín quarries. In the latter it was in the waste from silver mining and appeared alongside linarite and cerussite or with galena at a micro level.

Clinoclase $Cu_3AsO_4(OH)_3$ A hydrous copper arsenate mineral. Hardness 2.5 to 3. It is found in Mina Perdida in Cartagena and in the Sierra, in yellow and green forms.

Copiapite $Fe^{2+}Fe^{3+}_4(SO_4)_6(OH)_2 \cdot 20H_2O$ Hardnesss 2.5. Triclinic. It occurs as yellow micro crystals alongside Pyrite and Marcasite in the Los Blancos quarries. It is also found in the mines Tesoro, Belleza and Iberia.

Coronadite $Pb(Mn^{<4+},Mn^{2+})_8O_{16}$ It occurs as micro crystals alongside pyrolusite or barite in the Ramblizo de Las Nogueras and from Pepito and Herculano and in some grey stalactite masses in other mines of the area. It is also sometimes found alongside goethite.

Crocoite $PbCrO_4$. Lead chromate. Hardness 2.5 to 3. Monoclinic. Orangey red. There is a specimen in Madrid from Mina San Juan Bautista. There were two mines of this name and this is probably the one that was next to the San Valentín, quarry and which has now disappeared.

Cronstedtite $Fe^{2+}_2Fe^{3+}(Si,Fe^{3+}O_5)(OH)_4$ A rare iron silicate mineral. It occurs in Brunita alongside pyrite and pyrrhotite.

Deviline $CaCu_4(SO_4)_2(OH)_6.3H_2O$. Hardness 2.5. Monoclinic. Turquoise. Occurs as a micro in the San José and Gloria quarries. The colour is visible to the naked eye, but it can only be identified under a microscope. It can be confused with poor quality malachite from the same area, unless viewed under magnification.

Dickinsonite $KNa_4Ca(Mn^{2+},Fe^{2+})_{14}Al(PO_4)_{12}(OH)_2$ Hardness 3.5 to 4. Monoclinic. It appears alongside arrojadite in the Los Blancos quarry in masses of jarosite and natroalunite.

Dickite $Al_2Si_2O_5(OH)_4$ Hardness 1.5 to 2. Monoclinic. White. It occurs in the La Crisoleja area and Corta Sultana.

Dietrichite $(Zn,Fe,Mn)Al_2(SO_4)_4 22H2O)$ Hardness 2. Monoclinic. Dirty white. It occurs in microscopic state in Los Blancos near the road that runs from La Unión to Los Nietos

Dyscrasite Ag_3Sb A silver antimonide mineral. Hardness 3.5 to 4. Orthorhombic. It appears alongside carbonates of iron and copper in the Sierra de Cartagena and alongside galena and silver minerals in the San Valentín quarry.

Ecandrewsite $(Zn, Fe2+, Mn2+)TiO_3$ Hardness 5 to 5.5. Trigonal. Brown. Its qualities are similar to those of ilmenite. This mineral is not really collectable as it is only defined by analysis. Interestingly it is only found in two places in the world, Broken Hill and the San Valentín quarry.

Embolite $Ag(Br,Cl)$ Hardness 1.5 to 2. Isometric. It has been found in Mina Santa Bárbara and Mina Humo in La Unión. It also occurs with atacamite in tiny yellow or brown crystals on the barite found in the La Gaviota mine in Portmán.

Epsomite $MgSO_4.7H_2O$. Hardness 2. Orthorhombic. Often acicular. It is found in the mines Tesoro, Belleza, Catón, Dificultad and Iberia and in the Brunita quarry.

Ferrihydrite $(Fe^{3+})_2O_3 \cdot 0.5H_2O$ A powder-like nano mineral found in El Estrecho de San Ginés.

Fibroferrite. $Fe^{3+}(SO_4)(OH) \cdot 5H_2O$, Hardness 2.5. Hexagonal. It has been found chiefly in the slag from Mina San Lorenzo. It is fibrous and golden and appears in fissures. It was found after rocks from old galleries that had been flooded were extracted in 1983. It has also been found in the Brunita quarry.

Fosgenite $Pb_2(CO_3)Cl_2$ Hardness 2.5 to 3. Tetragonal. It has been found in the Cándido, Ferruginosa and Primitiva mines and in the San Valentín Quarry on goethite. It is present also in the Mar Menor due to material from early mining being dumped there.

Fraipontite $(Zn, Al)_3(Si,Al)_2O_5(OH)_4$ Hardness 2.5 to 4. Monoclinic. Green, yellow or blue. It occurs as a micro alongside hemimorphite and cerussite from the San Valentín quarry.

Glaucophane $Na_2(Mg_3Al_2)Si_8O_{22}(OH)_2$ Hardness 6 to 6.5. Monoclinic. Blue. occurs in micro specimens in areas of the Sierra where iron was mined.

Goslarite $(ZnSO_4 \cdot 7(H_2O))$ a hydrated zinc sulphate. Hardness 2 to 2.5. Orthorhombic. It is very unstable as it dehydrates to form other minerals. It occurs in Mina Amapola and in the Sierra on zinc blende from the Lolita, Julio César, Loba and Arresto mines.

Halotrichite $FeAl_2(SO_4)_4 \cdot 22H_2O$ A hydrated sulphate of aluminium and iron. It is water soluble. It is also known as feather alum. Hardness 1.5 to 2, monoclinic. White. it occurs in Brunita and the Aries, Numancia, Tesoro, Belleza, Cruz Chiquita and Iberia mines.

Halloysite $Al_2Si_2O_5(OH)_4$ Hardness 2 to 2.5. Monoclinic. It occurs in the La Crisoleja area, La Unión and in Corta Sultana.

Hawleyite CdS Cadmium sulphide. Hardness 2.5 to 3. Isometric. Orange. It has been found in the Los Blancos quarries.

Hinsdalite $PbAl_3(P_{0.5}S_{0.5}O_4)_2(OH)_6$ Hardness 4.5, Trigonal.. It occurs in micro form in Los Blancos and in the Cuesta de Las Lajas and in the Los Blancos quarry.

Hollandite $Ba(Mn^{4+}_6Mn^{3+}_2)O_{16}$ Hardness 4 to 6. Monoclinic. It is chocolate coloured and occurs in Mina Soledad in Portmán alongside barite.

Hydrohetaerolite $ZnMn_2O_4 H_2O$ Hardness 5 to 6. Tetragonal. Black. It occurs at a microscopic level alongside chalcophanite and the hemimorphite of Mina Precaución.

Hydrozincite $Zn_5(CO_3)_2(OH)_6$ Basic zinc carbonate. Hardness 2 to 2.5. Monoclinic. White. It occurs in minute quantities in mines with sphalerite. It occasionally covers specimens of sphalerite in Túnel José Maestre. It has appeared as small globules on quartz from the Emilia quarry and also in Corta San Valentín. It has been found in the Los Blancos quarries on galena and at a micro level in pieces from Mina Precaución alongside hemimorphite and calcite.

Kalinite $KAl(SO_4)_2 \cdot 11H_2O$ Hydrated potassium aluminium sulphate. Hardness 2 to 2.5. Monoclinic. White. It occurs in Cabeza Rajao.

Kaolinite $Al_4(OH)_8Si_4O_{10}$ Hardness 2 to 2.5. Triclinic. It occurs in the La Crisoleja area and in Corta Sultana.

Lepidocrocite γ-FeO(OH) Hardness 5. Orthorhombic. It is also known as esmeraldite or hydrohematite. It is found at micro level on Cabezo de San Ginés.

Linarite $PbCu(SO_4)_2$ Copper lead sulphate hydroxide. Hardness 2.5. Monoclinic. It occurs alongside hydrozincite, aurichalcite, caledonite and zinc blende in the Sierra de Cartagena. It also occurs in a brilliant blue-green colour alongside cerusite and anglesite in the walls of the San Valentín quarry.

Manganite MnO(OH) Manganese oxide hydroxide. Hardness 4. Monoclinic. Black. It occurs in the ramblas of Francés, de la Pocilgas, de Ponce, de Los Lobos and in El Abenque.

Mendozite $NaAl(SO_4)_2 \cdot 11H_2O$ A hydrated form of sodium aluminium sulphate. Hardness 3. Monoclinic. Colourless. It occurs in Cabezo Rajao.

Mimetite $Pb_5(AsO_4)_3Cl$ Hardness 3.5 to 4. Hexagonal. It has been found with Galena and Cerussite

Minnesotaite $(Fe^{2+},Mg)_3Si_4O_{10}(OH)_2$ Hardness 1.5 to 2. Triclinic. Greenish-gray. It occurs at a micro level alongside opal, chalcedony, magnesite and greenalite from the San Valentín, Emilia and Gloria quarries

Natroalunite $NaAl_3(SO_4)_2(OH)_6$ A sodium rich form of alunite. It is also called almerite. Hardness 3.5 to 4. Trigonal. It appears in the Los Blancos quarry with jarosite. It exists in micro form in the Cuesta de Las Lajas and in the Los Blancos quarry

Natrojarosite $NaFe^{3+}_3(SO_4)_2(OH)_6$ Hardness 2.5 to 3.5. Trigonal. It occurs in micro form in the Cuesta de Las Lajas and Los Blancos quarry

Nontronite $(CaO_{0.5},Na)_{0.3}Fe^{3+}_2(Si,Al)_4O_{10}(OH)_2 \cdot nH_2O$. Hardness 1.5 to 2. Monoclinic. It appears in San Valentín, the Collado de San Juan, Santa Isabel, Cuatro Santos and the barranco de Los Pajarillos and near Mina Catón.

Olivenite Cu_2AsO_4OH A copper arsenate mineral. Hardness 3. Monoclinic. Green. It occurs in limestone in Mina Consuelo, Cartagena and in La Verdad de Un Artista.

Pharmacolite $CaHAsO_4 \cdot 2(H_2O)$ A calcium arsenate. Hardness 2 to 2.5. Monoclinic. It is associated with Scorodite in the Sierra de Cartagena. Has high arsenic levels, so wash hands after handling.

Pickeringite $MgAl_2(SO_4)_4 \cdot 22(H_2O)$. It is a magnesium aluminium sulphate mineral. Hardness 1.5. Monoclinic. White. It occurs in the Brunita and Los Blancos quarries.

Plumbogummite $PbAl_3(SO_4)(PO_4)(OH)_6$ Hardness 4 to 5. Trigonal. Blue. It occurs in Cuesta de Las Lajas and in the Los Blancos quarry

Plumbojarosite $PbFe^{3+}_6(SO_4)_4(OH)_{12}$ Hardness 1.5 to 2 Trigonal. It occurs in Las Lajas and in the Los Blancos quarry.

Psilomelane $(Ba,H_2O)_2Mn_5O_{10}$ A manganese oxide sometimes wrongly called black hematite. Its primary constituent is romanechite. Hardness 5 to 6. Monoclinic. Blackish. It occurs in the Haití and Africana mines and the hillsides nearby and with other oxides in mines of the barranco de Las Nogueras. It is in stalactite form in the Africana mine.

Rutile TiO_2. Hardness 6 to 6.5. Tetragonal. It occurs in the beaches of Calblanque, Cuesta de Las Lajas and Los Blancos. Its colour is dark blue and it can be seen at a micro level alongside anatase, quartz, brookite and barite from Los Blancos.

Stephanite Ag_5SbS_4 A silver antimony sulphosalt mineral Hardness 2 to 2.5. Orthorhombic. Grey metallic. It occurs in the San Valentín quarry with other silver minerals and in La Unión.

Stibnite Sb_2S_3 Hardness 2. Orthorhombic. Metallic. It is also known as antimonite. It occurs in micro form with galena in many sites such as the mines San Rafael, Belleza, Paulina, Agradecida, Julio César, Rosa and Amable. It is on the Collado de Portmán, Sancti Spíritu and in the San Valentín, San José and Gloria quarries at a micro level. In the Rosa mine it is found with greenalite. It is also associated with pyrite in the Brunita quarry. It is also found in a mine on Cabezo de la Porpuz in Escombreras.

Svanbergite $SrAl_3(PO_4)(SO_4)(OH)_6$ Hardness 5. Trigonal. It occurs in micro form in Los Blancos, and Cuesta de Las Lajas.

Talc $Mg_3Si_4O_{10}(OH)_2$. Basic magnesium silicate. Hardness 1. Monoclinic. Often white. It occurs in the Brunita quarry and in the Huerto de San Pedro area close by on crystals of dolomite.

Tenorite CuO Copper oxide. Hardness 3-5 to 4. Monoclinic. Grey. It is also known as melaconite. It occurs in the Amable and Balsa mines in Cabezo del Pino.

Tetrahedrite $Cu_9Fe^{2+}_3Sb_4S_{13}$ A copper antominy sulphosalt. Hardness 3.5 to 4, It has been found with pyrrhotite in the Los Blancos quarry.

Tirolite $Ca_2Cu_9(AsO_4)_4(CO_3)(OH)_4 \cdot 11\text{-}12(H_2O)$ Hardness 1.5 to 2. Monoclinic. It occurs in the limestone of La Crisoleja.

Todorokite $(Na,Ca,K,Ba,Sr)_{1-x}(Mn,Mg,Al)_6O_{12} \cdot 3\text{-}4H_2O$. Hardness 1.5. Monoclinic. It occurs at micro level on Cabezo de San Ginés.

Vermiculite $(Mg,Ca,K,Fe^{11})_3(Si,AL,Fe^{111})_4O_{10}(OH)_2O_4H_2O$ Hardness 1 to 1.5. It occurs in the San Valentín and Tomasa quarries.

Witherite $BaCO_3$. Barium carbonate. Hardness 3 to 3.5. Orthorhombic. Cream colour. It occurs on barite in Mina Gaviota in Cola Caballo close to Portmán. There is a piece in the La Unión museum also from Mina Marinera.

Wulfenite $PbMoO_4$. Lead molybdate. Hardness 6.5 to 7. Triclinic. Orange. Wulfenite occurs in specimens of quartz and barite in yellow or orange colour.

Wolframite $(Fe,Mn)WO_4$ An iron manganese tungsten mineral. Hardness 4 to 4.5. Monoclinic. Grayish brown. It occurs rarely in the Sierra de Cartagena alongside other minerals such as quartz, cassiterite, arsenopyrite, etcetera.

Zincolivenite $CuZn(AsO_4)(OH)$ Hardness 3.5. Orthorhombic. Greenish blue. It occurs in Mina Consuelo.

31: Peñarroya chimney and pond

Bibliography

Minerales de la Región de Murcia by Maríano Muelas Espinosa, Pedro Pérez Nieto and Jordi Gil García-Miguel

Pió Wandosell Gil by Gonzalo Wandosell Fernández de Bobadilla

Patrimonio Cultural y Yacimientos de Empleo en la Sierra Minera de Cartagena-La Unión by a group attached to Fundación Sierra Minera

El Monasterio y Las Ermitas de San Ginés de la Jara by Alejandro Egea Vivancos

Las Ermitas del Cerro de San Ginés by Benjamín Mercader Sevilla

El Hundimiento del Castillo Olite by Luis Miguel Pérez Adán

*La Unión, El paisaje, el cante, el trovo, la min*a by Asensio Sáez

Libro de La Unión by Asensio Sáez

Crónicas del Festival Nacional del Cante de las Minas by Asensio Sáez

El Cante de las Minas by Rufo Martínez Cobacho

Letras de Cante (La Carpeta de Pencho Cros) Compiled by Fulgencio Cros

El País de la Plata by Ginés Pérez Garrigos

Crónica: Portmán Década de los 70 by Paco Baños Martínez

Le Zinc by Jacques Duchaussoy

Bocamina Volumen 2

Garum Sociorum: La Industria de salazones de pescado en la Edad Antigua en Cartagena by Javier R. García del Toro

El Minero Romano de Carthago Nova Vestimenta e instrumental by Javier R. García del Toro.

Centro Minero de la Sociedad Minera y Metalúrgica de Peñarroya España, S.A. en Cartagena La Unión.

Sucedió en Cartagena by José Luis Sánchez Álvarez

Minerales de la Unión by Ginés López García

Rothschild by Miguel A. López-Morell

La Quema de la Asamblea by Manuel Ponce Sánchez.

Washing Amethysts in the Bidet

32: Túnel José Maestre behind reeds

Fiona Pitt-kethley

Index

A
Acanthite 278
Adamine 278
Adamite 126, 278
Agave 41, 151
Águilas 12, 231
Aguirre 58, 230, 231
Alabaster 128
Alcohol De Vidriero 117
Alex De La Iglesia 194
Alfonso García 243, 244
Alfonso Martínez Saura 33
Algameca 54, 159, 229, 231
Alhama La Seca 73, 232
Almadraba 85, 197
Almerite 283
Alum 53, 55, 91, 158, 159, 261, 282
Alumbres 60, 79, 135, 137, 158-68, 231, 234, 261, 263, 264, 266, 267, 275
Aluminacopiapite 278
Aluminium 175, 278, 280, 282-284
Alunite 283, 261
Amblygonite 278
Amethyst 4, 39-41, 43, 46, 181, 182, 209, 215, 261, 266, 273
Ammonal 212
Ana María Céspedes Soler 74
Ana Rama 4, 68
Anatase 278, 280, 284
Andalusians 16
Andalusite 278
Andrés Moreno García 125
Andrés Pedreño Torralba 231
Andrés Teulón 234

Andrew Wood 4
Ángel Miralles 162
Anglesite 91, 209, 211, 215, 262, 265, 275, 279, 283
Ankerite 266, 279
ANSE 86, 218, 256, 257
Antimonite 262
Antimony 279, 284
Antonio "Sevillano" 4
Antonio Grau Dauset 30
Antonio Grau Mora 30
Antonio Muñoz 75
Antonio Pagan Lorenzo 112
Antonio Retamero Tejada 4
Aragonite 20, 46, 47, 262, 264
ARBA 162, 218
Archaeological Museum 146
Archaeologists 15, 19, 151, 176
Argentojarosite 279
Armada 217
ARQUA 124
Arrojadite 279
Arsenic 92, 263, 284
Arsenopyrite 215, 285, 263
Arturo Perez Reverte 124
Asensio Sáez 25, 75, 287
Asoc Cultural Mineralógica de la Sierra de Cartagena-La Unión 4, 18
Asturias 81
Atacamite 281, 279
Atamaría 63, 94, 95, 97, 201
Audouin gull 121
Aurichalcite 279
Azurite 126, 209, 211, 263, 266, 271

B
Badajoz 235
Baebelo 87
Banco de Cartagena 234
Barite 21, 96, 97, 125, 137-139, 164,
　　165, 182, 187-189, 191, 213,
　　263, 264, 266, 268, 271, 273,
　　274, 278-282, 284, 285
Bar Minero 24
Barranco Bilbao 44
Barranco de los Pajarillos 47
Barranco de Mendoza 61
Barranco Feo 135
Bartolomé Spottorno 231
Barus 137
Batería de la Parajola 122
Batería de los Dolores 168
Batería Doctrinal de Brigadas 60
Bazan 244
Befesa 162
Berber 147, 159
Bertrandite 279
Beudantite 279
BIC 248-250
Birnessite 269, 279
Bismuth 225
Bituminaria Bituminosa 176
Bocamina 18, 24, 288
Bodega Velasco 163
Bornite 264
Botryogen 279
Boulangerite 279
Bournonite 266, 279
Braunite 269, 279
Broken Hill 210, 281
Bronze 13, 52
Brookite 280
Burnt Amethyst 43, 266

C
Cabezo Agudo 57
Cabezo de Don Juan 207, 265, 278
Cabezo de la Porpuz 168, 284
Cabezo del Pino 187, 201, 271, 285
Cabezo de Ponce 61, 94, 263, 265,
　　283, 288
Cabezo de San Ginés 61, 262, 264,
　　268, 269, 273, 274, 279, 280,
　　283, 285
Cabezo Rajao 51, 54, 55, 57, 59, 60,
　　63, 64, 81, 200, 235, 261, 267,
　　268, 270-272, 275, 276, 283
Cabo de Palos 31, 80, 96, 117, 149,
　　269, 274

Cadmium 15, 173, 179, 246
Café Cantante 29, 30
Cala Cortina 68, 128, 129, 164, 200,
　　269
Calamine 61, 141, 174, 269, 275
Cala Reona 117
Calblanque 12, 63, 95, 96, 98, 115,
　　137, 138, 159, 245, 269, 271,
　　278, 284
Calcite 20, 21, 48, 96, 97, 125, 142,
　　182, 203, 216, 262, 264-266,
　　273, 283
Calcium 262, 264, 266
Caledonite 280
Californios 130
Callosa de Segura 30
Calvario 130, 133, 134
Calzada Romana 201
Camilo Aguirre 231
Camilo Calamari 168
Campos de Golf 94, 95, 138, 166, 263,
　　264, 266, 273, 274
Cante de las Minas 16, 21, 27, 31, 32,
　　56, 69, 75, 113, 194, 287
Cantera Tomasa 100, 209, 210
Capa Negra 211, 215
Carbonates 54, 59, 215, 281
Carboneras 186
Caridad La Negra 236
Carlos Lanzarote map 130
Carrots Café 31
Cartagenera 234, 266
Carthagena & Herrerías Steam Tram-
　　way Company Ltd 60, 78
Carthaginians 15, 84, 124, 147
Casa Cervantes 13
Casa del Folclore 31
Casa del Piñon 27, 31
Casa del Pueblo 88
Casa de Misericordia 14
Casas-Cañada 83
Casas de Emiliano 83
Casas de Espín 83
Casciario 79
Cassiterite 105, 264
Catedral de Cante 27, 31
Caustic Soda 221
CEAM 226
Celdrán 47, 63, 109, 141, 188, 212,
　　224, 231, 232
Celestine 200, 210, 265
Celestino Bonifacio Martínez 234
Celestino Martínez 232, 235

Centro de Estudios y Analisis Mineralurgicos 226
Centro Minero de la Sociedad Minera y Metalurgica de Peñarroya-España, S.A 224
Cerussite 209, 262, 265, 273, 280, 282, 283
Cervantite 280
Chalcanthite 280
Chalcedony 105, 209, 215, 265, 272, 275, 283
Chalcocite 280
Chalcophanite 142, 280, 282
Chalcopyrite 129, 188, 206, 213, 215, 257, 265, 274
Chamosite 280
Chlorargyrite 278, 280
Chlorite 215
Choncholita 186
Chronicle of Al Himyari 145
Chrysocola 215
Cinnabar 265
Citrine 42, 43, 266
Civil War 9, 58, 62, 75, 80, 109, 122, 168, 236
Clinoclase 280
C. Numisius 146
Cocotazos 215, 231
Codex Calixtinos 145
Cola de Caballo 187, 188, 202, 264, 265, 285
Colectividad Minera CNT-UGT 58
Collado de las Tinajas 47
Compagnie du Chemin de Fer de la Sierra de Carthagene 79
Compagnie Française des Mines et Usines d'Escombrera-Bleyberg 44, 109, 231
Compañía Cartagenera de Navegación 234, 235
Compañía de Portmán 79
Compañía Minera Bético Manchega 215
Concesión Brandt 187
Concesión Sivet 187
Control de Leyes 225
Copiapite 280
Copper 15, 52, 53, 61, 62, 68, 81, 126, 129, 145, 173, 175, 205, 221, 228, 229, 242, 263-266, 275, 276, 279, 280, 281, 283-285
Copper Sulphate 53, 221
Coronadite 280
Corta Blancos 90

Corta Brunita 62, 63, 113, 169, 193, 201, 206, 210-215, 221, 263-267, 270-276, 279-285
Corta Emilia 43, 59, 63, 172, 179, 183, 206-209, 211, 215, 220, 230, 262, 265, 267, 268, 270, 272, 275, 283
Corta Gloria 12, 59, 63, 205, 206, 211, 212, 215, 220, 224-265, 268, 269, 271, 273, 275, 279, 281, 283, 284
Corta Sultana 94, 206, 215, 216, 221, 265, 267, 270, 272, 273, 275, 280, 281, 282, 283
CreeCT 162, 218
Crocoite 281
Cronstedtite 213, 281
Cuesta de Las Lajas 61, 111, 112, 262, 264, 270-272, 278, 279, 282-85
Cuevas de Roma 79
Cueva Victoria 143, 144, 147, 262
Cuprite 266
Curro Piñana 194
Cyanide 173, 221

D

DANA 252
Deviline 205, 206, 211, 212, 260, 281
Dickinsonite 279, 281
Dickite 281
Diego Martínez de Ojeda Martínez 4
Dietrichite 281
D. Manuel Rodríguez Esparza 125
Dolomite 213, 266, 285
Dolores Calvache Yáñez 233
Domingo Jiménez Campillo 55, 58
Don Juan Conesa 125
Dos Hermanos 73, 233, 234
Duke of Escalona 158
Dyscrasite 278, 281

E

Ebonite 167
ECAndrewsite 209, 210, 281
El Abrevadero 79
El Algar 147, 235, 262
El Beal 69, 87, 255
El Carmoli 278
El Chicharra 78
El Chorrillo 112
El Chupa 89
El Cónsul 33

El Descargador 5, 61, 67-72, 78, 79, 87, 103, 155, 207, 209, 233
Eleuterio Andréu Martínez 56
El Garbanzal 10, 53, 54, 70, 231
El Gorguel 6, 54, 58, 63, 124, 126, 139, 159, 160, 172, 186, 188, 191-193, 196-198, 231, 250, 263, 267, 268, 271, 276
Eliécer Pérez Sánchez 4, 163, 164
El Infierno Prometido 193
El Lazareto 61
El Lirio 94, 95, 232
Eloy Celdrán 63, 212, 231
El Pais 226
El Paraje de La Finca del Pajarillo 44
El Rico 158
El Sabinar 83, 278
Embolite 279, 281
En El Tranvía 58
Enrique Carrión Inglés 58, 208, 268
Epsomite 281
Eremita de Los Ángeles 149
Escombrera-Bleyberg 44, 46
Escombreras 60, 61, 83, 109, 120-126, 128-130, 134, 135, 141, 161, 163-165, 168, 191, 198, 212, 230, 238, 265, 267, 284
Escombreras Chamomile 121
Española del Zinc, S.A. 207
Esparto Grass 51, 159
Estanislao Rolandi Bienest 79
Esteban Llagostera 231
Estrecho de San Ginés 80, 83, 94, 139, 141, 235, 263, 273, 279, 282
Explosivos Rio Tinto S.A. 224

F
Fábrica de Pío 74
Fernando Gómez 133
Ferrihydrite 282
Ferrocarriles Españoles de Via Estrecha 80
Ferruginosa 274, 282
FEVE 70, 72, 77, 78, 80, 81, 86, 167, 236, 254, 268
Fibroferrite. 282
Figueroa 78, 79, 231
Flooding 56, 59, 61, 91, 179, 213
Fluorite 96, 137, 158, 163-167, 210, 263, 266, 267, 279
Fluorophosphate 278
Fool's Gold 53, 113
Fosgenite 282

Fraipontite 282
Francés 267, 273, 283
Francisca Calvache Yañez 233
Francisco Carrillo Paredes 74
Francisco Celdrán Conesa 212
Francisco Dorda Lloveras 55, 231
Francisco J. Ródenas Rozas 75
Francisco Molero Rubio 74
Franco Citti 194
Franco-Española 230
Fray Melchor 147, 148
Fuensanta Alcaraz Saura 4
Fuente Alamo 122, 123, 161
Fulgencio Martínez Conesa 55
Fundación Pura 79
Fundación Vedruna 71
Fundición Dos Hermanos 73
Fundición Pío Wandosell 73
Fundición Santa Lucia 224
Fundación Sierra Minera 25, 287

G
Gabriel Gutiérrez Sánchez 74
Gaceta Minera y Comercial de Cartagena 59
Galena 55, 58, 59, 61, 62, 78, 104, 115-118, 125, 137, 182, 187-189, 209, 211, 213, 215, 224-226, 261-263, 267, 268, 273-275, 278-281, 283, 284
Garum 83-86, 124, 288
General Requeña 54
Gilberte Florentine 34
Ginéa Sanz Giménez 74
Ginés El Franco 149
Ginés García Millán 194
Ginés López García 4, 288
Gitanos 9, 70
Glaucophane 282
Goethite 104, 206, 209, 211, 216, 267, 273, 281, 282
Gold 15, 52, 53, 81, 113, 173, 228, 229, 268
Goslarite 282
Gran Hotel 75, 235
Greenalite 59, 116, 211, 263, 264, 267, 268, 270, 276, 279, 283, 284
Greenockite 268, 276
Greenpeace 63, 172, 175, 246
Green Quartz 43, 187
Greta Thunberg 254
Guardia Civil 168
Gypsum 20, 128, 187, 188, 200-203, 213, 269

H
Halite 269
Halloysite 282
Halotrichite 278, 282
Hansa Urbana 145, 146, 256
Hares 41
Hawleyite 282
Headframe 57, 72, 73, 99, 103, 126, 187, 188, 198
Hematite 129, 269, 284
Hemimorphite 20, 141, 142, 187, 269, 270, 282, 283
Herminio Añón Martínez 74
Hermitage 150-152, 154, 155, 161
Hilarión Roux 44, 60, 109, 121, 141, 148, 230, 231
Hinsdalite 282
Hollandite 282
Hospital Minero 74
Huerto de San Pedro 168, 266, 285
Huerto Pío 35, 67, 68, 233
Hydrohetaerolite 282
Hydroheterite 142
Hydrozincite 142, 283

I
Iglesia de Caridad 146
Ignacio de Figueroa 78
Ignacio Figueroa y Mendieta 231
Ilmenite 210, 281
Iron 13, 15, 27, 52, 53, 60-62, 79, 96, 98, 102, 108, 110, 126, 129, 141, 167, 187, 188, 208, 224, 231, 261-263, 265, 267, 269, 270- 276, 280-282, 285
Iron Hydroxide 267
Isabelle Carbonell 4
Island of Escombreras 83

J
Jarosite 270, 279, 281, 283
Jaspe 102, 271
Javier García del Toro 4, 84, 144, 153
Javier Lorente, 154
Joaquín Peñalver Nieto 27, 231
Joaquín Toscano 192
José De Luis Del Campo 4
José Javier Saura Nadal 4
José López Martínez 235
José Luis Guirao Escudero 4
José Maestre 12, 63, 64, 110, 178, 183, 210, 220, 222, 231, 236, 238, 261, 263, 264, 267, 268, 273-275, 278, 283

José Matías Peñas 4, 91
Juan José Martínez Pardo, 4
Juan Manuel Chumilla 193, 196
Juan Pinilla 75
Juan Robles 69

K
Kalinite 283
Kaolinite 283
KSK 237

L
La Amistad 233
La Aparecida 146, 147
La Artesiana 57
La Buena Muerte 147
La Caridad Hospitalaria 233
La Coruna 133
La Crisis 10, 115, 253
La Crisoleja 233, 237, 265, 267, 272, 275, 280-283, 285
La Española de Zinc S.A 224
La Esperanza 54, 60, 62, 63, 79, 80, 160, 168, 195, 214, 229, 235, 264
La Familia 233
La Manga 105, 145, 172, 246, 252, 254, 257
La Palma 31
La Parreta 60, 62, 160, 167, 168, 274, 275
Larim 148
Las Cenizas 231
Las Mateas 83
Las Matildes 67, 69, 248
Las Nogueras 280, 284
La Sociedad Los Intransigentes 57
Laures 148, 149
Laurinum 148
Lavadero Roberto 63, 64, 100, 173, 175, 177, 178, 187, 208-210, 220-222, 224-226
La Verdad 126, 131, 155, 198, 263, 278, 284
Lead 15, 46, 52, 53, 55, 60-62, 73, 98, 99, 115-118, 121, 129, 135, 137, 159, 169, 173-175, 186, 187, 189, 208, 224, 226, 228, 230-232, 237, 243, 244, 246, 262, 265, 267, 268, 270, 273, 279, 281, 283, 285
Leandro 149
Legón 29, 43, 249
Le Havre 60

Lepidocrocite 283
Ley de Minas de Fausto de Elhuyar 53
LIFE 99, 218
Liliane 212
Limonite 59, 139, 141, 142, 207, 263, 269, 270, 274
Linarite 266, 280, 283
Lithopone 175
Llamusi 25
Llano del Beal 12, 41, 43, 44, 59, 61-63, 72, 74, 79, 83, 86-88, 90, 94, 96, 97, 115, 116, 118, 139, 141, 155, 164, 172, 181, 208, 216, 232, 234, 236, 243, 255, 261, 264, 266, 268, 273, 276
Lo Campano 79, 128, 131-135
Loma del Engarbo 187
Loma del Teniente 280
Lo Pagan 253
Lo Poyo 86, 256
Lorca 89, 231, 239, 276
Los Belones 95, 278
Los Blancos 61, 78-80, 83, 94, 206, 214, 215, 261, 262, 265, 267, 268, 270-273, 275, 276, 278-285
Los Caballeros de Santa Bárbara 122
Los Camachos 35
Los Cucones 264, 269
Los Nietos 78-81, 83, 86, 141, 254, 258, 281
Los Nietos Nuevos 83
Los Partidarios 167
Los Quebrarados 272
Los Urrutias 80
Los Villares 83
Ludlamite 213, 270, 273, 276

M

Magnesite 283, 270
Magnesium 175, 266, 284, 285
Magnetite 59, 270
Malacates 61
Malachite 112, 126, 129, 205, 211, 263, 266, 271, 281
Mancommunidad de Heraderos de los Dorda 55
Manganese 15, 61, 79, 96, 137, 141, 187, 231, 261, 268, 280, 284, 285
Manganite, 269, 283
Manto Piritoso 215
Manuel Morales 4, 24, 223

Manuel Rodríguez Gil 233
Manuel Rodríguez Wandosell 233
Maquinistas de Levante 32, 236
Marcasite 125, 215, 263, 264, 271, 280
Mar de Cristal 253
Mares Bravas 128
Margarita Lozano 194
María Cegarra Salcedo 179, 180, 181
María del Carmen Hevia de Saavedra 122
María de los Ángeles 55
María Dolores 46, 230, 262, 274
Maríano Roca 243, 244
María Piedad Aguirre Aldayturriaga 230
María Visitación Zapata 236
Marmatite 271
Mar Menor 85, 86, 91, 244, 252, 253-257, 260, 282
Marqués de Villamejor 78, 231
Marquis of Vélez and Molina 158
Marrajos 130
Martinete 56, 194, 195
Mazarrón 12, 51, 52, 73, 158, 231, 233, 236
Media Legua 54
Melaconite 285
Melanterite 271
Mendoza 59, 61, 79, 90, 267, 271, 276
Mendoza Cargadero 79
Mendozite 283
Mercury 20, 52, 53, 265, 271
MetalEurop, S.A 226, 243
Meteorite 19
Methane 144
Miguel Hernández 181
Miguel Zapata Sáez 44, 61, 160, 178, 231, 234, 236-238
Mimetite 283
Mina Abundancia 215
Mina Afortunada 208
Mina Africana Segunda 231
Mina Agradecida 263, 266, 284
Mina Agrupa Vicenta 69, 105, 108, 109, 171, 240, 272
Mina Amable 212, 271, 284, 285
Mina Amapola 282
Mina Amigos Consecuentes 230
Mina Amistad 187
Mina Ángelita 213
Mina Anticipada 230
Mina Arresto 271
Mina Aries 62, 213, 271, 282
Mina Artesiana 57, 200, 269, 272, 275

Mina A Santelvas 198
Mina Balsa Depositaria 200-202, 269
Mina Bella Unión 208
Mina Belleza 72, 208, 263, 265, 266, 272, 275, 279-282, 284
Mina Bilbao 59
Mina Blanca 67, 262
Mina Braguelona 200, 230, 269
Mina Brigida 44
Mina Buena Suerte 211, 212
Mina Buen Consejo 266, 271, 275, 279
Mina Camarón 124, 198
Mina Candelaria 99
Mina Cándido 282
Mina Carlos 24, 53, 55, 96, 130, 179, 215
Mina Carlos y Emilia 179
Mina Carolina La Doncella 274, 275
Mina Carpeana 53, 138, 263
Mina Catón 44, 230, 272, 281, 284
Mina Colmenera 230
Mina Cometa Donati 231
Mina Conchita 208
Mina Concilio-Consuelo 198
Mina Constancía 212
Mina Constancía de un Amigo 230
Mina Consuelo 271, 278, 284, 285
Mina Crescencia Segunda, 230
Mina Cruz Chiquita 62, 271, 280, 282
Mina Cuarenta y Cinco 215
Mina Cuarta 264
Mina Cuatro Santos 230, 284
Mina de las Palomas 96
Mina Descuidado 179, 208
Mina Depositaria 187
Mina Dichosa 212, 230, 275
Mina Dicido 213
Mina Dificultad 200, 269
Mina Dios Te Ampare 198
Mina Don Carlos 55
Mina Dos Amigos 198
Minados La Pobrecita 44
Mina Ebraldo 231
Mina Edetana 231
Mina El Ángel de la Guarda 230
Mina El Arresto 58, 267, 275
Mina El Cielo 73
Mina El Concilio 62
Mina El Corcho 230, 265, 275
Mina El Español 231
Mina El Espectador 111
Mina El Huerto 275

Mina El Humo 265, 269
Mina El Juanito 208, 213
Mina Eloisa 230
Mina El Submarino 96
Mina El Tábano 230
Mina El Tesoro 110, 271
Mina El Tranvía 57, 58
Mina Emilia 208
Mina Emma 231
Mina Encarnación 212
Mina Encontrada 230
Mina Enrique VIII 208
Mina Esperanza 264, 274
Mina Felisa 208
Mina Feliz Anuncia 187
Mina Fortuna 52, 103, 265
Mina Frasquita 208
Mina Fulgencio 44, 46, 55, 149, 287
Mina Gaviota 265, 269, 285
Mina Grandeza 230
Mina Guadalupe 167
Mina Haití 263, 264, 268
Mina Herculano 96, 97, 231, 264, 273, 280, 281
Mina Herrera 179
Mina Humboldt 189, 201
Mina Humo 281
Mina Iberia 55, 56, 200, 230, 269, 272, 275, 276, 279-282
Mina Inglesa 111, 263, 265
Mina Isabelita 231
Mina Isabel la Católica 208, 230
Mina Jacinta 103
Mina Jenny 90, 91, 215, 232
Mina Jesualda 215
Mina Joaquina 234, 273, 274
Mina Jorge 138
Mina Josefita 208
Mina Julio César 232, 266, 271, 275, 276, 279, 282, 284
Mina Júpiter 208, 230
Mina La Africana 213
Mina La Buscada 215
Mina Laberinto 198
Mina La Carolina 275
Mina La Carpeana 138, 263
Mina La Chapinas 230
Mina La Conciliación 96
Mina La Gaviota 187, 202, 264, 279, 281
Mina La Higuera 96
Mina La Ligera 208
Mina La Loba 275
Mina La Loca del Capellán 230

Mina La Lucera 208
Mina La Mona 215
Mina La Montañesa 167
Mina La Murciana 275
Mina La Ocasión 57, 58, 62
Mina La Pagana 44
Mina La Paloma 230
Mina La Paz 233
Mina La Pura 72
Mina La Purisima Concepción 236
Mina La Suerte 231
Mina La Superior 265
Mina La Tonta 244
Mina La Verdad de Un Artista 126, 278
Mina León Negro 49, 208, 230
Mina Loba 99
Mina Lola 268, 274, 275
Mina Lolita 62, 267, 275, 282
Mina Los Aragonitos 47, 262
Mina Los Burros 213
Mina Los Negros 265
Mina Los Pajaritos/Pajarillos 46, 232, 284
Mina Lo Veremos 59, 275
Mina Lucera 61, 208, 230, 237
Mina Lucifer 213
Mina Madrileñita 208
Mina Manolita 59, 96, 97, 167, 264
Mina Manto de los Azules 59, 62, 105, 206, 211
Mina María Jesus 55
Mina Mariana 208
Mina Marisol 96, 97, 166, 262, 263, 273
Mina Marinera 231, 264, 285
Mina Marquesita Moderna 280
Mina Mas Alerta 99
Mina Mendigorría 44, 62
Mina Mentor 44
Mina Mercurio 274
Mina Monserrat 57-57
Mina Neptuno 79
Mina Niño Jesus 141
Mina No, No 130
Mina Nunca Vista 230
Mina Obdulia 139, 188
Mina Observación 230
Mina Observación a Santelvas 198
Mina Ocasión 63
Mina Olivares 79
Mina Olvidada 231
Mina Oportunidad 198, 233
Mina Oriolana 230

Mina Pablo y Virginia 109, 110, 230, 271
Mina Paulina 103, 263, 266, 284
Mina Pepito 262, 280
Mina Perdida 230, 275, 280
Mina Permuta 188
Mina Plutón 231
Mina Pobrecita 44, 230
Mina Por Si Acaso 230
Mina Porvenir 230
Mina Pozo Cuevas 72
Mina Precaución 141, 143, 231, 264, 269, 270, 280, 282, 283
Mina Presentación 187
Mina Previsión 230
Mina Primera Paz 90
Mina Primitiva 282
Mina Rafaela 215
Mina Reforma 230
Mina Reina Regente 124
Mina Remunerada 264, , 271
Mina Reserva 230
Mina Revolución 57, 58, 208, 230
Mina Rómulo 48, 117, 230, 262, 267
Mina San Águstin 128, 269
Mina San Aniceto 140, 141, 226, 231
Mina San Antonio 153, 262, 265, 268, 274, 279
Mina San Bartolome 187
Mina San Benito 213
Mina San Bruno 231
Mina San Camilo 136, 137, 164, 167, 258, 261, 263, 266, 274, 279
Mina San Dionisio 99
Mina San Eloy 230
Mina San Francisco Javier 198, 211
Mina San Isidoro 55, 61, 149, 208, 264, 276
Mina San Jacinto 208
Mina San Jerónimo 187 Mina San Joaquín 230
Mina San Jorge 72, 230, 263
Mina San José 12, 59, 63, 172, 202, 205, 206, 211, 212, 215, 224, 230, 265, 268, 270-273, 275, 281, 284
Mina San Juan Bautista 60, 67, 208, 263, 265, 266, 281
Mina San Luciano 47
Mina San Manuel 230
Mina San Nicolás 215
Mina San Pedro 21, 146, 168, 197, 208, 214, 223, 253, 266, 285
Mina San Quintin 67

Mina San Rafael 63, 125, 126, 198,
    230, 272, 274, 275, 279, 284
Mina San Roque 160, 161, 166
Mina San Sebastián 79, 215
Mina San Simón 167, 230
Mina Santa Adelaida 53
Mina Santa Ana y San Juan 230
Mina Santa Antonieta 69, 125, 198
Mina Santa Bárbara 122, 139, 191,
    263, 281
Mina Santa Catalina 55
Mina Santa Eduvigis 44
Mina Santa Filomena 230
Mina Santa Florentina 31, 149, 208
Mina Santa Isabel 198, 262, 284
Mina Santa Leocadia 61
Mina Santa Teresa Salvadora 212
Mina Santo Tomás 99
Mina San Timoteo 139, 189, 201,
    263, 269, 280
Mina Satanás 213
Mina Secretaría 98, 99, 274, 275
Mina Segundo Ferrocarril 117, 187,
    188, 200, 223, 267, 269
Mina Telémaco 67
Mina Santa Teresa 212, 214, 230,
    272, 275
Mina Segunda Esmeralda 272
Mina Segunda Paz 62, 90, 208, 232
Mina Segundo San Rafael 125
Mina Sin Duda 230
Mina Si Puede Ser 230
Mina Si, Si 130
Mina Suerte y Verdad 212
Mina Tercera España 265
Mina Tercera Española 215
Mina Tercera Esperanza 280
Mina Teresita. 138, 263, 273, 280
Mina Tesoro 110, 271, 279, 280,
    281, 282
Mina Tetuán 275
Mina Tiburón 130
Mina Usurpación 208
Mina Usurpada 208
Mina Valarino 78
Mina Venus 208
Mina Vibora 274
Mina Virgen de los Ángeles 208
Mina Victoria 264, 274
Mina Vigilante 215, 230
Mina Violeta 230
Mina Virgen de las Mercedes 215,
    231
Mina Virgen del Carmen 230

Mina Virgen de los Ángeles 208, 230
Mina Virgen del Pilar, 230
Mina Virgen de Monserrat 55
Mina Vulcano 231, 265
Mina Yo Que Se 162
Mina Zurbano 230
Minas Cartes 63
Minas de Alcohol 117
Minas de Cortes, S.A. 55, 58
Minas de Pinturas 112
Minera Celdrán 47, 63, 109, 141, 212,
    224
Minerales San Juan 63, 125
Mineral Fair 18, 22, 27, 69, 96, 108,
    113, 116, 137, 201
Minera Navidad 63
Mineras 16, 27, 28
Mineros de Verdad 126
Mínguez Inglés 158
Minnesotaite 283
Molienda Semiautógena 209
Mompean Bridge 78
Monasterio de San Ginés 145, 256
Monheimite 209, 271, 275
Monoclinic 263, 268-271, 273, 276,
    278-285
Monte Calvario 130, 133
Monte Carmelo 215
Monte de las Cenizas 189, 245
Monte del Pino 187
Monte Laberinto 139, 188
Monte Miral 141, 143, 145, 147, 149
Monte Roldán 149
Monte Sacro 64, 84, 166, 250
Mr. Witt 31
MU88 172
MURAM 109
Murder 33, 34, 37, 131, 134

N
N332 54, 137, 167
Nación Española 179, 208
NASA 214
National Archaeology Museum 51
Natroalunite 279, 281, 283
Natrojarosite 283
Nitrates 254
Nitro-glycerine 167
Nontronite 284
No te Escaparás 230
NPK fertiliser 255
Nueva California 130
Nueva Santa Lucia 132
Numancia 282

O

Ochre 102, 104, 112, 159, 271, 273
Olivenite 284
Opal 20, 102-105, 116, 209, 215, 265, 267, 271, 272, 283
Open-Pit 62, 63, 113, 206
Orcelitana 53, 187, 236
Orcharson and Enthoven 125
Oro de Tontos 113
Oxalic Acid 165

P

Palacio Consistorial in Cartagena 88
Palacio Pedreño 79
Paraje de Las Pocilgas 264
Paraje de la Torrecica 57
Paraje de Quebrarados 264
Pedreño-Aznar 79
Pedro Cerdán Martinez 27
Pedro Conesa 231
Pedro García Valdés 75
Peña de Águila 94, 97, 98, 189, 210, 245, 274
Peña de Antonio Piñana 181
Peñarroya 4, 6, 56, 60, 62, 64, 87, 109, 112, 118, 121, 125, 129, 175, 176, 178, 179, 212, 220, 223-226, 239, 243, 244, 245, 267, 271
Pencho Cros 28, 287
Petrochemical Industry 212
Pharmacolite 284
Phoenicians 15, 84, 124, 147
Pickeringite 278, 284
Pilar de Jaravía 200
Piñon 27, 31
Pío Wandosell 15, 59, 67, 72-74, 231, 232, 234, 287
Piritas de Suerte 113
Playa de los Nietos 83
Playa Flamenca 24
Playa Honda 83, 253
Plaza de Rey 30, 31
Plaza de San Francisco 178
Pliny 40, 86, 87
Plumbogummite 284
Plumbojarosite 284
Poblado Repsol 124, 193
Polybius 87
Porcelainitas 272
Portmán Golf 56, 63, 64, 87, 90, 109, 118, 138, 154, 175, 176, 214, 223, 239, 243-246
Portmaur 231

Portus Magnus 11, 172
Pozo Carmelo 58
Pozo Mercurio 179, 183
Prasiolite 43, 272, 273
Preexplotación 225
Preussag, A.G. 226
Professor del Toro 84
Pronta 230
Psicofonias 147
Psilomelane 274, 284
Puertos de Murcia 30
Punic Wars 15
Pyrite 14, 53, 58-60, 62, 69, 79, 104-106, 108-113, 125, 171, 207, 211-213, 215, 217, 221, 224, 226, 257, 263, 264, 267, 270-273, 276, 281, 284
Pyrolusite 273, 274, 280
Pyromorphite 273
Pyrrhotite 263, 272, 273, 276, 281, 285

Q

Quartz 20, 21, 32, 39, 40-44, 47, 48, 102-104, 110, 117, 125, 165, 167, 182, 187, 206, 209, 211, 213, 264-268, 270-273, 274, 275, 278, 280, 283-285
Queen Isobel II 72

R

Rambla de Crisoleja 187
Rambla de las Colmenas 100, 139, 189
Rambla de las Nogueras. 97
Rambla del Avenque 60, 198, 202
Rambla de Mendoza 59, 90, 271, 276
Rambla la Boltada 60
Ramblizo de las Nogueras 263, 264, 268, 280
Ramón Ramirez 160
Real Compañía Asturiana de Minas 56
Real Madrid 16, 233
Recyclex, S.A. 226
Red Ochre 104
Regente Group. 62
RENFE 81
Repsol 124, 161, 162, 193, 223
Rhodochrosite 98, 274
Ricardo Guardiola 59
Rincón de San Ginés 83
Rito y Geografía 56
Robert Barnes 4
Roche 10, 11, 54, 146, 160, 161
Rodalquilar 158
Rogelio Mouzo Pagán 4, 112, 125

Rojo El Alpargatero 29
Roland 149
Roldán 54, 122, 149
Roman 15, 47, 52, 53, 54, 72, 83, 84,
    85, 103, 111, 121, 144, 146,
    147, 148, 149, 154, 178, 189,
    243, 252
Romanechite 284, 274
Rosa 232, 263, 284
Rothschild 46, 54, 60, 175, 230, 288
Ruston Bucyris 214, 216
Ruta 33 105, 109, 171, 172, 206, 208,
    211, 267
Rutile 280, 284

S
Saint Antonio 150
Saint Ginés 231
Saint Hilarión 150
Saint Jerome 150
Saint Paul the Hermit 150
Salida de Operarios de Don Miguel
    Zapata en La Unión 238
Salinas 86, 223, 257, 269
Salvadora 212, 230
Sancti Spiritu 54, 60, 61, 62, 70, 72,
    79, 98, 103, 104, 105, 106,
    116, 200, 217, 231, 265, 266,
    267, 269, 270, 272, 273,
    275, 284
Sanctuario de Los Ángeles 151, 153
San Domingo 161
San Felix 168
San Ginés 52, 61, 83, 141, 145, 146,
    150, 152, 233, 235, 256, 262,
    263, 264, 268, 269, 273, 274,
    279, 280, 282, 283, 285, 287
San Ginés de la Jara 83, 145, 287
San Ignacio 118, 167
San Ignacio foundry 118
San Julian 118, 120, 128-130, 269
San Julian Castle 130
San Lorenzo 57, 58, 64, 282
San Onofrio 150, 151
Santa Lucia 30, 60, 62, 68, 78, 79,
    80, 118, 130, 132, 135, 207,
    224, 229, 230, 244, 253, 267,
    271, 278
San Valentín 12, 59, 63, 116, 179,
    182, 206, 208-211, 217, 220,
    222, 224- 226, 261-285
San Vicente del Raspeig 19, 21, 116
School of Mines 59
Scorodite 274, 284

Selenite 128
Señor Ortega 112
Serafín Cervantes 231
Serpentine chimney 110, 111
Sevillano 4, 155, 164
Siderite 11, 125, 182, 206, 209, 211,
    213, 215, 264, 268, 270, 274,
    276
Sierra de Almagrera 12, 186
Sierra Gorda 135, 168
Siervas de Jesús 74
Silicate(s) 59, 108, 206, 268, 269, 274,
    280, 281, 285
Silicon Dioxide 261, 265, 266, 272,
    273
Silver 15, 52, 53, 55, 60, 61, 78, 87,
    105, 115, 116, 117, 148, 152,
    171, 173, 228, 230, 232, 246,
    268, 275, 280, 281, 284
Slate 13, 110, 111, 171
Smithsonite 141, 211, 213, 267, 268,
    269, 271, 275
SMMP 207
SMMPE 63, 212, 213, 214, 223
Snakes 41
Soc. Anónima Minera Los Pobres 141
Soc. Anónimo Civil La Alternativa
    125
Soc. de Explosivos 168
Soc. Franco Española de Explosivos
    234
Soc. Minera y Metalúrgica de
    Peñarroya-España 56, 87,
    207, 288
Soc. Minero Metalúrgica Zapata-
    Portmán 62
Soc. San Fulgencio 44, 46
Soc. San Hilarión 109
Société Minière et Métallurgique de
    Peñarroya 46, 60
Specularite 129, 269
Sphalerite 182, 268, 271, 274-276, 283
Stalactites 182, 188, 216, 262, 264, 268
Stannite 215, 276
Stephanite 278, 284
Stibnite, 168, 262, 284
St. James 253
St. Jean of Arles 145
St. Lucy 48
Strabo 55
Sulphantimonite 279
Sulphates 54, 59, 225
Sulphide. 21, 263, 265, 267, 271, 272,
    273, 275, 279, 282

Sulphur 108, 109, 110, 111, 188, 214,
    215, 268, 276
Sulphuric Acid 120, 217, 221, 224
Svanbergite 285

T
Tabano 208
Talc 285
Talía 233
Tapería Edward 70
Tartessans 15, 52, 84, 147
Tenorite 285
Teodosia 149
Teresa Rosique 65
Tetraclinis Articulata 218
Tetrahedrite 215, 285
The Castillo Olite 122
The Fundación of Santa Lucia 207
Tin 15, 62, 105, 264, 276
Tío Lobo 160, 178, 186, 236, 238
Tirolite 285
Titanium 210
Todorokite 269, 285
Tomasa 59, 63, 100, 206, 207, 209,
    210, 221, 224, 262, 264, 265,
    267, 268, 269, 270, 271, 272,
    275, 285
Tomas Rico 235
Torremendo 230
Torrente 230
Torre Pacheco 34, 254
Torrevieja 24, 165
Toscanini 192
Tragsa 90
Tres Hermanas 233
Triclinic 278, 280, 283, 285
Trincabotijas 130
Trinidad 55, 111, 265, 275
Trituración Primaría 209
Túnel José Maestre 12, 63, 178, 220,
    222, 261, 263, 264, 267, 268,
    273, 274, 275, 283

U
Unión Española de Explosivos 61,
    212
University of Murcia 59, 176

V
Valentín Escobar Callejón 74
Venta El Descargador. 69
Ventorillo Guirao 70
Vera 12
Vermiculite 285

Victor Beltrí 13, 27, 88, 133, 178,
    235, 237
Villa 42 97
Villa Dolores 67, 233
Villa Paris 80
Virginio Moreno 18, 138
Vista Alegre 135, 137, 160, 167, 263,
    266, 279
Vivianite 113, 213, 270, 273, 276

W
Wandosell y Toledano 233
Wells 16, 40, 56, 57, 58, 90, 134, 153,
    160, 163, 166, 179, 256
Windmill 104, 105, 160
Witherite 285
Wolframite 285
Wormwood 41
Wulfenite 285

X
Xanthates 221

Y
Yeso 200

Z
Zamak 175
Zinc 13, 15, 21, 53, 55, 59-62, 173-175,
    181, 187, 207, 209, 211, 221,
    224, 228, 243, 244, 246, 261,
    264, 267, 269, 273, 275-279,
    282, 283, 288
Zinc Blende 21, 60, 173, 181, 209,
    261, 264, 267, 273, 275, 279,
    282, 283
Zincolivenite 285
Zincsa 244
Zinc sulphate 173

Figure Copyrights

1: Cala Cortina **8** © author

2: La Unión Market **17** © author

3: Tunnel interior, La Unión **23** © herraez - stock.adobe.com

4: La Unión Town Hall **26** © SoniaBonet stock.adobe.com

5: Washing Amethysts in the Bidet **38** based on a photo © myper - stock.adobe.com

6: Mina Catón **45** © author

7: View of Calblanque and Cabo de Palos © Werner Wilmes

8: El Descargador **66** © author

9: The FEVE **76** © author

10: Ermita de San Onofre Monte Miral **82** © author

11: El Lirio **93** © author

12: Opals **101** © author

13: Pyrite **107** © VV Voennyy - stock.adobe.com

13: Galena **114** © jonnysek - stock.adobe.com

14: Escombreras Industry **119** © author

15: Stones from the temple of Júpiter Stator on Monte San Julian **127** © author

16: Barite **136** © author

17: Barite, Cartagena **140** © Bergminer CC Wiki Media

18: Fluorite **157** © jirivaclavek - stock.adobe.com

19: Sphalerite **170** aka zinc blende, blende, blackjack, mock lead © AC stock.adobe.com

20: Antiguo lavadero de mineral Portmán **185** © cherokeerose stock.adobe.com

21: Beautiful El Gorguel beach in Cartagena Province, Spain **190** © SoniaBonet - stock.adobe.com

22: Gypsum, Cartagena **199** © CC Wiki Media

23: Microscope **204** © Thorsten Katz - stock.adobe.com

24: Lavadero Roberto Portmán **219** © cherokeerose - stock.adobe.com

25: Gran Hotel **227** © author

26: The end of mining lavadero roberto plus Portmán beach **241** © author

27: Calcination Oven, El Gorguel **247** © author

28: Mar Menor **251** © author

29: Calcination Oven near El Lirio **259** © author

30: El Laberinto **276** © author

31: Peñarroya chimney and pond **286** © author

32: José Maestre behind reeds **289** © author

Mineral Images
Pages 261 - 276
© *stock.adobe.com*
© *CC Wiki Media*
© *author*

Milton Keynes UK
Ingram Content Group UK Ltd.
UKHW050004051223
433799UK00001B/6